POLYETHERS AND POLYETHYLENE GLYCOL

CHARACTERIZATION, PROPERTIES AND APPLICATIONS

Chemical Engineering Methods and Technology

Additional books in this series can be found on Nova's website
under the Series tab.

Additional e-books in this series can be found on Nova's website
under the e-book tab.

CHEMICAL ENGINEERING METHODS AND TECHNOLOGY

POLYETHERS AND POLYETHYLENE GLYCOL

CHARACTERIZATION, PROPERTIES AND APPLICATIONS

JOYCE WILLIAMS
EDITOR

New York

NOTICE TO THE READER

Library of Congress Cataloging-in-Publication Data

Polyethers and polyethylene glycol : characterization, properties and applications / Joyce Williams, editor.
    pages cm. -- (Chemical engineering methods and technology)
  Includes bibliographical references and index.
  ISBN 978-1-63483-392-9 (softcover)
  1. Polyethylene glycol. 2. Nanostructured materials. I. Williams, Joyce, 1972- editor. II. Cong, Hailin.
Preparation and application of photosensitive polyethylene glycol in micro-nanofabrication.
  TP248.65.P58P64 2014
  668.4'23--dc23
                        2015025925

Published by Nova Science Publishers, Inc. † New York

CONTENTS

PREFACE

The authors of this book examine polyethers and polyethylene glycol. The first chapter in particular gives a systematic, balanced and comprehensive summary of the main aspects of photosensitive polyethylene glycol. The second chapter is a focus on the biodegradation of polyethylene glycols (PEGs) and polyethoxylated surfactants. The third and last chapter of this book focuses on polyethylene glycol based phase change polymers for thermal energy storage applications. Life nowadays is often centered around technologies, and it is of utmost important to have the primary and secondary resources ready for present and future needs. This last chapter addresses these concerns and various procedures in which thermal energy can be stored and used.

Chapter 1 - Photosensitive polyethylene glycol (PEG) which is obtained through modification of PEG has been on use for a few decades and is well established among many fields. Modified PEG is known for its longer chains, better solubility; contemporary enhance bioavailability, contributed to resist proteolytic degradation and reduce immunogenicity. Due to having excellent anti-protein-fouling property and biocompatibility, the research on PEG has surprisingly blossomed during recent decades, and many practical micro-nano materials and biochips based on photosensitive PEG with potential applications in drug delivery, protein separation, cell culture and biomedical engineering have been generated. In this chapter, we give a systematic, balanced and comprehensive summary of the main aspects of photosensitive PEG which related to its modification methodologies, structure, properties and the potential applications driven in biological research as micro-nano materials and biochips, drug delivery, precipitating agent for proteins, biomedical engineering which will be discussed in the meantime.

Chapter 2 - Research on biodegradation of polyethylene glycols (PEGs) and polyethoxylated surfactants has a long history of over 50 years. Aerobic and anaerobic microbial degradation of PEGss and polyethoxylated surfactants has been documented, although the mechanism of anaerobic degradation has not been fully elucidated. It is noteworthy that ether bonds are resistant for hydrolysis except acidic conditions, which are not the cases for most of expected biodegradation. PEG and polyethoxy chains are depolymerized step by step as one glycol unit from the hydroxyl terminals under either aerobic or anaerobic conditions. The aerobic metabolism of PEGs and polyethoxylated compounds is started by PEG dehydrogenase (PEG-DH). The *peg* operon was comprised of six genes including *pegA* (the gene encoding PEG-DH) in Sphingomonads. As *pegA* is likely to have been distributed among different PEG-degrading bacterial species, the *peg* operon or genes relevant to the metabolism of PEG are probably existent in a variety of bacteria. On the other hand, wood-rot fungi catalyze non-metabolic depolymerization of PEG *via* peroxidase or Fenton reaction, which happens at random on the polyethoxy chain. Thus biodegradation of PEGs and polyethoxylated compounds is no more of a big problem. PEGs appear to be metabolically inert and nontoxic, but some report described that `PEGs were sulfated *in vitro* by the rat and guinea pig livers, and carboxylated *in vivo* in rabbits and burn patients. In addition, as some depolymerized metabolites were reported as toxic and less biodegradable than original compounds, pursuit of metabolites of PEGs and polyethoxylates either in the environment or *in vivo* is meaningful.

Chapter 3 - Phase change material can make a great contribution to the energy saving by way of energy storage in the form of latent heat. It can act both ways as when temperature is to be reduced it melts from crystalline phase while temperature of the substrate is prevented from rising. Similarly, during heating, a pre-molten phase of a crystalline material releases the heat while getting crystallized. In the process further cooling of the substrate is prevented.

A series of strategies are incorporated in the chapter and their heat storage details are discussed. As the phase change material is polyethylene glycol (PEG) it has either to be used by constraining in microcapsule or to be prevented from flowing during melting by adopting other methodologies for convenient use.

The first strategy discussed is the form-stable composition which does not allow PEG to flow during melting due to various secondary forces restricting their separation. This strategy is practiced by physically mixing PEG with other polymers or porous carbons, silica, nanoclay and others. The process advantage is further enhanced by mixing some materials which enhance the thermal conductivity ensuring fast action.

Chemically reacted PEG copolymers and others are always non-flowable although the enthalpy may be sacrificed.

PCM fibers and textiles are also discussed. In both form stable, microcapsulated PCM adherence to textile as well as direct chemical linking of PCM with textile have been discussed.

Lastly, the incorporation in concrete, building structure, PU foam etc. are discussed and their advantages and disadvantages are discussed for better understanding.

The characterization and performance study are also incorporated in the chapter.

In: Polyethers and Polyethylene Glycol   ISBN: 978-1-63483-392-9
Editor: Joyce Williams   © 2015 Nova Science Publishers, Inc.

Chapter 1

PREPARATION AND APPLICATION OF PHOTOSENSITIVE POLYETHYLENE GLYCOL IN MICRO-NANOFABRICATION

Hailin Cong[,1,2], Bing Yu[#,1,2], Xiaodan Xu[1], Xin Chen[1] and Hua Yuan[1]*

[1]College of Chemical Engineering,
Qingdao University, Qingdao, China
[2]Laboratory for New Fiber Materials and Modern Textile,
Growing Base for State Key Laboratory,
Qingdao University, Qingdao, China

ABSTRACT

Photosensitive polyethylene glycol (PEG) which is obtained through modification of PEG has been on use for a few decades and is well established among many fields. Modified PEG is known for its longer chains, better solubility; contemporary enhance bioavailability, contributed to resist proteolytic degradation and reduce immunogenicity. Due to having excellent anti-protein-fouling property and biocompatibility, the research on PEG has surprisingly blossomed during recent decades, and many practical micro-nano materials and biochips based on photosensitive PEG with potential applications in drug delivery, protein

---

[*] hailincong@yahoo.com
[#] Email: yubingqdu@yahoo.com

separation, cell culture and biomedical engineering have been generated. In this chapter, we give a systematic, balanced and comprehensive summary of the main aspects of photosensitive PEG which related to its modification methodologies, structure, properties and the potential applications driven in biological research as micro-nano materials and biochips, drug delivery, precipitating agent for proteins, biomedical engineering which will be discussed in the meantime.

Keywords: polyethylene glycol, photosensitive, anti-protein-fouling, stability, biochips, biomedical engineering

1. INTRODUCTION

Polyethylene glycol (PEG) is neutral, hydrophilic polymers which is also soluble in all kinds of organic solvents. Once in aqueous solution they are heavily hydrated, highly mobile and resist other polymers (including proteins and nucleic acids). Molecules coupled to PEG become nontoxic, nonimmunogenic, soluble in water and a lot of organic solvents, and surfaces modified by PEG-coating attachment become hydrophilic and resistance protein adsorption [1]. These properties have gave rise to a great diversity of biotechnical and biomedical applications containing: aqueous two-phase partitioning [2-4], biomolecule immobilization [5-9], drug modification [10-13], and preparation of protein rejecting surfaces [14-17]. Additionally, photosensitive PEG-coated surfaces can be used to control wetting and electroosmosis [18-20].

In order to achieve more desired properties of PEG, a series of photosensitive polyethylene glycol introduced by physical or chemical modification in the past decades. Since PEG groups have been used extensively in biological and pharmaceutical applications to improve the pharmacokinetic properties of biomolecules [21-23]. PEG bioconjugates provide enhanced in vivo stability and are nontoxic and nonimmunogenic [24-27]. Hydrogels, due to characteristic properties such as swellability in water, hydrophilicity, biocompatibility, and lack of toxicity, have been utilized in a wide range of biological and medical applications [28-33]. A novel nonionic hydrogel, polyethylene glycol-cinnamylidene acetate (b-PEG-CA) can be described which is photosynthesis, photoscission, and preliminary blood biocompatibility [34]. Photosensitive poly(ethylene glycol) (PEG) hydrogels, due to characteristic properties such as swellability in water, hydrophilicity and lack of toxicity, have been used in a wide range of biomaterials including a delivery carrier and tissue engineering. They have been prepared by cross-linking PEG with a ultraviolet (UV) irradiation [35, 36] or a peroxide [37], photopolymerization of acrylate and methacrylate attached to PEG initiated by visible light or long wave UV [38, 39]. Most of these photocrosslinking have been performed in the presence of photoinitiators and/or photosensitizers. PEG hydrogel is also prepared by photodimerization of photosensitive groups such as cinnamilidene acetate or anthracene which was added as terminal functional groups to PEG [40–43]. Additionally, a novel PEG surface modification and patterning process has been proposed [44, 45]. This approach mimics traditional photolithography in that spin-coating and UV exposure through a photomask are employed to create a polymer pattern. However, rather than form a barrier to penetration of etching agents, PEG hydrogel microstructures resist protein or cell adhesion [46, 47]. These modifications of PEG all have excellent photosensitive performance and a wider range of application than the pure PEG. In this chapter, we give a systematic, balanced and comprehensive summary of the main aspects of photosensitive PEG which related to its modification methodologies, structure, properties, and the potential applications driven in biological research as micro-nano materials and biochips, drug delivery, precipitating agent for proteins, biomedical engine-eering which will be discussed in the meantime.

2. MODIFICATION METHODOLOGIES OF PEG

The first synthesis of polyethylene glycol (PEG) is attributed to the French chemist and Sorbonne professor Charles Adolph Wurtz. He derived ethylene oxide in 1859 by the action of caustic alkali on ethylene chlorohydrins [48]. Then Wurtz reacted ethylene oxide with water to derive ethylene glycol. He obtained the lower-molecular-weight polyethylene glycols along with the ethylene glycol, and those polyethylene glycols first gained technical importance [48]. And then, many modification mothods of PEG have been researched by scholars.

2.1. Photocleavable PEG Reagent

The chemical modification reactions on the preformed polymers can be divided conveniently into three types such as crosslinking, chemical combination with different polymers and structural modification of the polymer chain which includes copolymerization, block and graft.

The bulk of the PEG polymer wraps around the protein, thus providing the desired protective effects by sterically blocking it without perturbing the conformation and dynamics of the protein native fold [49, 50]. For this reason, PEGs containing an ester linkage for tethering to the biomolecule have been developed that are susceptible to spontaneous hydrolysis and provide a slow release pathway for protein delivery [51, 52]. While hydrolyzable linkers facilitate general release of the active therapeutic over time, a means to obtain precise spatial and temporal control over bioavailability without compromising stability or cellular uptake would be greatly advantageous.

Due to the interest in the photochemical control of biological processes [53-60], such control over protein function using light is sought to accomplish. Light affords a noninvasive tool for modulating biological function, as it can be modified in amplitude, duration, and location to achieve a high level of spatiotemporal resolution. It is reasoned that the installation of a photolabile PEG directly onto a protein would provide a PEG photocage that can inhibit protein activity (Figure 1A). Upon irradiation with non-phototoxic UV light, the PEG photocage will be removed from the protein, thus restoring its native activity. Since UV irradiation can be precisely controlled with high spatial and temporal resolution, this methodology provides a fundamentally new approach to achieve spatiotemporal control over protein activity. Herein, Georgianna et al. describe the first photocleavable PEG reagent (PhotoPEG). In addition, they demonstrate that it provides photochemical control over the activity of a PhotoPEGylated enzyme [27].

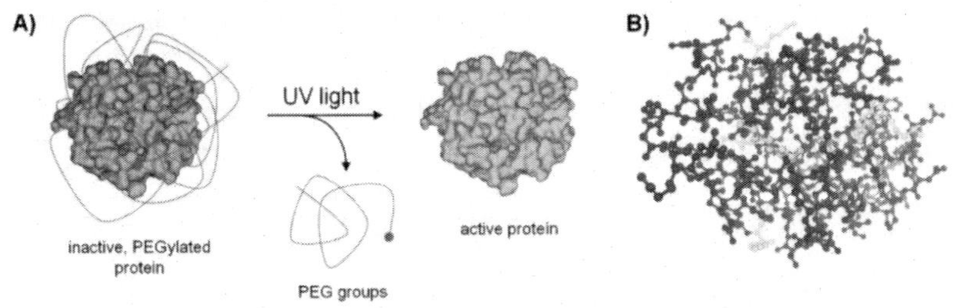

Figure 1. (A) Light-induced removal of PEG groups enables photochemical control of protein function. (B) X-ray structure of lysozyme with the six solvent-accessible lysine residues highlighted in yellow (PDB 2VB1) [27].

In order to demonstrate the applicability of our approach, we selected lysozyme as a target. Lysozyme is a 14.7 kDa protein that catalyzes the hydrolysis of 1,4-linkages in peptidoglycans and chitodextrins, thus destroying the cell walls of bacteria and protecting tissue from bacterial invasion [61]. For this reason, lysozyme has been extensively studied [62-64], its crystal structure has been solved [65], and assays to assess its enzymatic activity have been developed [66, 67]. Importantly, in addition to its N-terminus, lysozyme contains six solvent-accessible lysine residues (Figure 1B) and is thus readily PEGylated, which has previously been shown to inhibit its catalytic activity [67].

2.2. Hydrogel of PEG

Photopolymerization of b-PEG-CA was performed in a film cast from a dichloromethane solution on a cover slip. Figure 2 shows changes in the absorption spectra for b-PEG-CA (λ=313 nm) during increasing time of exposure to light. Spectral changes are indirect measurements of the disappearance of the double bond adjacent to the carbonyl group of the cinnamylidene acetate moiety and, at the same time, the formation of a cyclobutane ring. The sharp isosbestic point at 265 nm indicates that a single reaction is proceeding in the system, namely, the disappearance of the double bond of the cinnamylidene acetate moiety and the formation of crosslinks via cyclobutane formation [34]. As mentioned earlier, the cinnamylidene acetyl moiety undergoes a 2 + 2 cycloaddition reaction [68-73], which, in solution, transforms *trans*-cinnamylidene acetyl groups to *cis*-cinnamylidene acetyl groups upon exposure to light. The *cis*-form undergoes photodimerization via cyclobutane ring formation [68-72]. As shown in Figure 2, an 87% yield is accomplished in less than 10 min, in the complete absence of photoinitiators. Photogelation of photosensitive PEGs which were cast onto glass slides occurred as fast as 5 min. The gel point is defined as the point at which *b*-PEG-CA monomers upon irradiation with UV light formed an insoluble three-dimensional network in water.

Figure 3 illustrates the process flow for the fabrication of PEG hydrogel micropatterns. PEG hydrogel patterns were fabricated from the precursor solution of PEG-DA (MW 575) with 1% (w/v) photoinitiator, DMPA. This solution was spun at 1000 rpm for 6 s onto the silane-treated glass surface containing terminal acrylate orvinyl functional groups using a spin-coater (Machine World, Inc. Redding, CA). The uniform layer of the PEG-DA precursor solution on glass was then exposed through a chrome/soda lime photomask (Advance Reproductions, North Andover, MA) to 365 nm, 15 mW/cm^2 UV light from a Q 2001 mask aligner (Quintel Co., San Jose, CA). The exposure times ranged from 1 to 2 s.

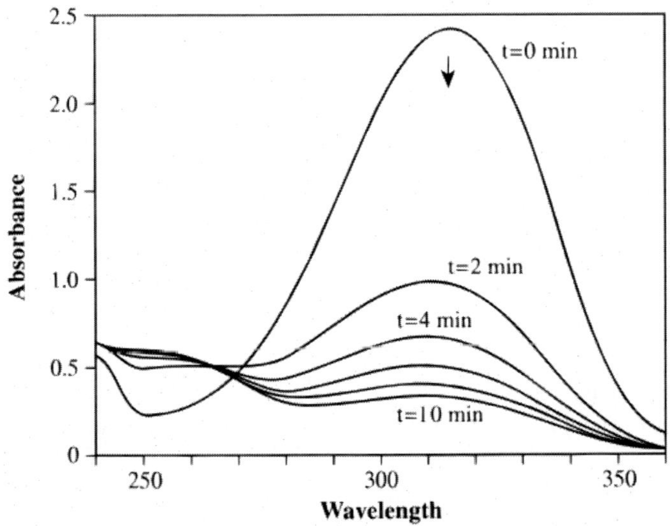

Figure 2. Change in absorption spectra of b-PEG-CA in water (2 mM) during irradiation (10 min) with a 150 W Xenon arc lamp [34].

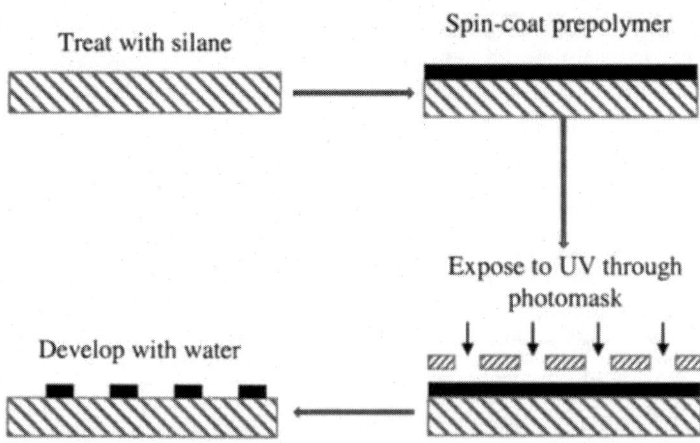

Figure 3. Process flow for the fabrication of PEG hydrogel microstructures [47].

The regions of PEG-DA exposed to UV light underwent free-radical polymerization and became cross-linked, while unexposed regions were dissolved in DI water after 5 min of development. The height of the resultant hydrogel microstructures varied from 2 to 5 µm, as measured with a Dektak3 surface profiler (Veeco Instruments, Santa Barbara, CA). Hydrogel microstructures for high-density cell patterning consisted of either an 80 × 80 array of wells with individual well dimensions of 30 × 30 µm and 20 µm walls or a 100 × 100 microwell array with 20 × 20 µm well dimensions and 20 µm walls [47].

2.3. Formation of Monodisperse PEG Microspheres

The PEG prepolymer was prepared bymixing the PEG-DA with 2 wt% 2-hydroxy-2-methyl-propiophenone. As shown in Figure 4, the PEG prepolymer was pumped into the silicon oil in the T-junction area, resulting in the formation of the uniform PEG prepolymer microspheres dispersed in silicon oil. The formed PEG prepolymer microspheres were collected and photopolymerized under 365 nm UV light at an intensity of 610 µW/cm^2 for 30 min, and the monodisperse PEG microspheres were obtained. The morphology and diameter of the obtained microspheres were characterized by an optical microscope (SMZ-T2).

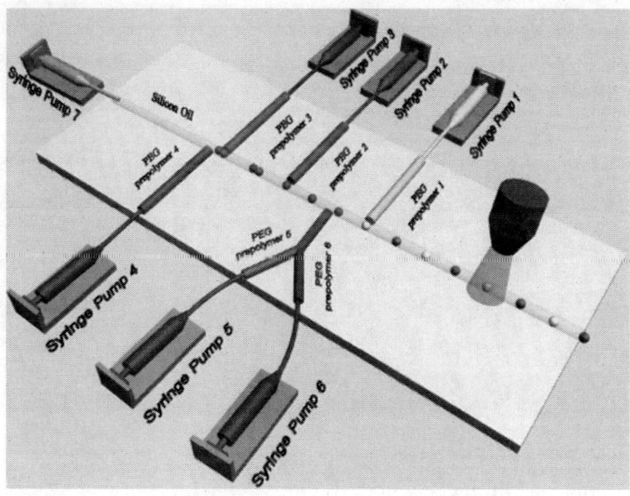

Figure 4. Schematic illustration of the fabrication process of monodisperse PEG microspheres [75].

As shown in Figure 5(*a*), peanut-shaped binary PEG/dye prepolymer emulsion droplets are formed in the microfluidic channel using the PEG prepolymer-3 and prepolymer-4 as the adjacent incontinuous sample phases. After online UV photopolymerization, the binary PEG/dye prepolymer droplets are crosslinked and merged to form peanut-shaped microparticles with different colors on each side (Figure 5 (*b*)).

As shown in Figure 5 (*c*), binary Janus PEG/dye composite prepolymer emulsion droplets are formed in the microfluidic channel using the mixed laminar flow of PEG prepolymer-5 and prepolymer-6 as the discontinuous sample phase.

After online UV photopolymerization, the binary PEG/dye composite prepolymer droplets are crosslinked to form binary Janus microspheres loaded with different colors synchronously (Figure 5 (*d*)). The selection of dyes with different affinities, e.g., one is oleophylic and the other is hydrophilic, is very important to the success of the microfabrication. Otherwise, the two colors will be merged and disappeared rapidly under the shearing force in the flow [75].

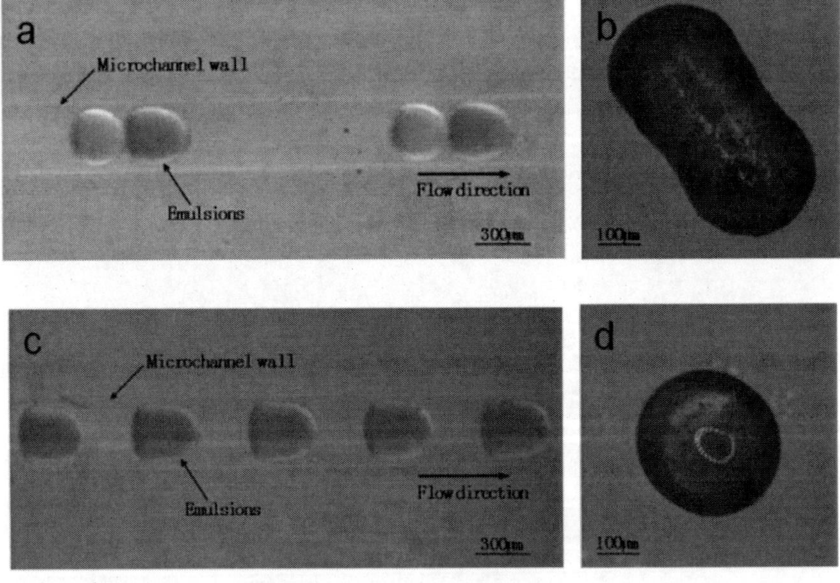

Figure 5. Photos of binary PEG/dye composite droplets and microspheres: (*a*) peanut-shaped binary emulsion droplets in microfluidic channel; (*b*) peanut-shaped binary microparticles after UV crosslinking; (*c*) binary Janus emulsion droplets in microfluidic channel; (*d*) binary Janus microspheres after UV crosslinking [75].

3. APPLICATIONS OF MICRO-NANO MATERIALS AND BIOCHIPS BASED ON PEG

Due to having excellent anti-protein-fouling property and biocompatibility, the research on photosensitive polyethylene glycol (PEG) has surprisingly blossomed during recent decades, and many practical micro-nano materials and biochips based on photosensitive PEG with potential applications in drug delivery, protein separation, cell culture, biomedical engineering and bio-membrane have been generated.

3.1. Application in Drug Delivery

Hydrogel microspheres are a kind of polymeric microparticles that have a hydrophilic nature and can take water in aqueous environments. PEG hydrogel is one of the crosslinked polyether that has characteristics including good biocompatibility, non-toxic ability and low immunogenicity in vivo and it is widely used in biomedical fields such as drug delivery. Monodisperse PEG hydrogel composite microspheres with different morphologies and composition are fabricated by microfluidic chip and UV photopolymerization system by our group [74, 75]. Preparation of monodisperse PEG microspheres by a T-junction microchip [74] and T-junction microchip equipped with rounded channels [75] are demonstrated, respectively. The effects of microfabracation conditions and composition components to the shape, size and properties of the obtained microspheres are preliminarily studied and discussed. The obtained PEG/aspirin composite microspheres exhibit a sustained release of aspirin for a wide time range.

Ma et al. demonstrated for the first time the ability to coat solid dispersions on microneedles as a means to deliver water-insoluble drugs through the skin [76]. PEG was selected as the hydrophilic matrix, and lidocaine base was selected as the model hydrophobic drug to create the solid dispersion. The microneedles coated with PEG-lidocaine dispersions resulted in significantly higher delivery of lidocaine in just 3 min compared with 1h topical application of 0.15g EMLA, the process offered a practical approach to coat water-insoluble drugs on microneedles for transdermal delivery. Dornianiand et al. immobilized biological materials on iron oxide nanoparticles' surface by coating it with PEG to minimize the aggregation of FNPs and enhance the effect of nanoparticles for biological applications [77].

The FNPs-PEG was loaded with perindopril erbumine (PE), an antihypertensive compound to form a new nanocomposite (FPEGPE). The PE loading (10.3%) and the release profiles from FPGP nanocomposite were estimated, and the result showed that up to 60.8 and 83.1% of the adsorbed drug was released in 4223 and 1231 min at pH 7.4 and 4.8, respectively. Cytotoxicity study showed that free PE presented slightly higher toxicity than the FNPs and FPEGPE nanocomposite. Hydrogels composed of gellan gum (CG) and polyethylene glycol dimethacrylate (PEG-DMA) were realized to overcome the fragility problems of physical gels of GG by Pacelli and co-workers [78]. Vitamin B12 and myoglobin were chosen to carry out release studies in order to have insight on the mesh size of the internal work of different hydrogels. The presence of PEG-DMA$_{750}$ was able to slow down the diffusion rate of the model protein, and this can be related to the lower ability of the matrix to swell in PB pH 7.4 with respect to SN of GG. Cui et al. conducted pharmacokinetic experiments in rates in order to determine the effect of interfacial PEG on the blood circulation lifetime and the biodistribution of PA/Cholesterol/PEG-Cholesterol liposomes [79]. Doxorubicin could be actively loaded in liposomes with a high drug loading efficiency and a high drug to lipid ratio. The lifetime of these non-phospholipid liposomes in the blood circulation was considerably shorter than that observed for control PEGylated phospholipid liposomes, a phenomenon associated with the negative interfacial charge of the PA/Chol/PEG-Chol liposomes. The microparticles made by Tamilvanan et al. showed a drug release profile that was completely in a retarded style when compared to the release profile of drug alone [80]. The antimicrobial activity of azithromycin (AZM) was not affected after incorporating it into microparticles as shown in the zone-inhibition assay. Nevertheless, the microparticles reduced markedly the red blood cells (RBC) rupturing property of the drug in comparison to drug in phosphate buffer solution of pH 7.4. This indicated that the microparticles prepared based on PEG could be of a suitable carrier to protect AZM from thermodegradation to provide retardation in drug release. PEG is used in various pharmaceutical preparations. Kinetics of two active pharmaceutical ingredients, cetirizine and indomethacin processing carboxylic acid functionality, has been studied in PEG 400 and PEG 1000 at different temperature by Schou-Pedersen and co-workers [81]. The shelf-life for cetirizine in a PEG formulation at 25°C expressed at t$_{95\%}$ was predicted to be only 30h. Further, rate constants for esterification of cetirizine in PEG 1000 in relation to PEG 400 decreased by a factor of 10. This drug excipient interaction can reduce the shelf-life of a low-average molecular weight PEG formulation considerably. Polymeric nanoparticles are beginning to make significant impact on global pharmaceutical sector. They are promising vehicles for site specific and controlled delivery of therapeutic agents. Moorkoth et al. aimed to synthesis nanoparticles of polylactic acid (PLA) and polyhydroxy butyrate (PHB) and their PEG combinations of desired size and to check their cytotoxicity and tolerance to the simulated biological fluids in order to explore them for drug attached oral applications [82]. PEG plays a key role in the stability of the nanoparticles in the GI fluids and in the access of the antigen to the blood and lymphatic circulation. Nanoparticles with PEG surface coating or PLA-PEG-PLA copolymer were successful in terms of stability at simulated gastric and intestinal fluids as well as cytotoxicity test. Thus, it is of primarily importance to validate the colloidal properties of nanoparticles exposed to the biological fluids.

3.2. Application in Protein Separation

The modification of proteins by covalent attachment of PEG can eliminate some drawbacks of native proteins and improve their physicochemical, biomedical and pharmacological properties [83].

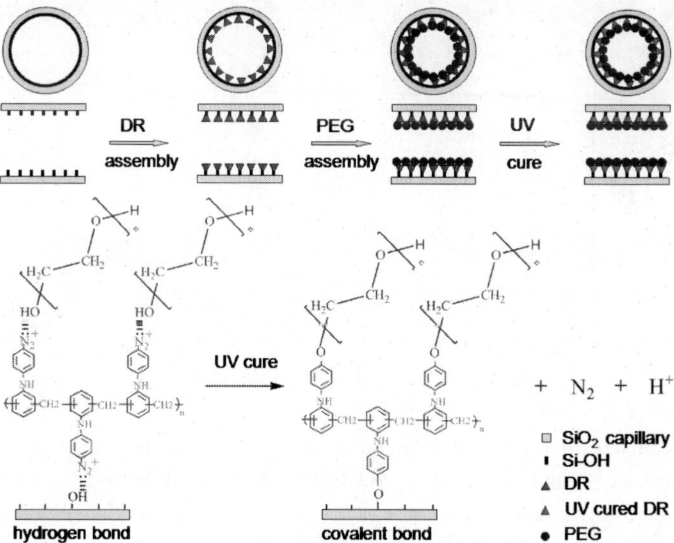

Figure 6. Schematic illustration of the preparation of DR/PEG covalent coating on capillary surface [84].

Numerous PEG-modified proteins show an extended plasma half-life, reduced antigenicity and immunogenicity, increased solubility and resistance to proteolysis, while often maintaining a high percentage of the biological activity. And for now, preparation of PEG-modified microchannels for protein analysis has been paid more attention to. In our works [84], a novel method for the preparation of covalently linked capillary coatings of PEG was demonstrated using photosensitive diazoresin (DR) as the coupling agent. Layer by layer (LBL) self-assembly film of DR and PEG based on hydrogen bonding was first fabricated on the inner wall of capillary, then the hydrogen bonding was converted into covalent bonding after treatment with UV light through a unique photochemistry reaction of DR, the schematic illustration of the preparation of DR/PEG covalent coating on capillary surface is shown in Figure 6.

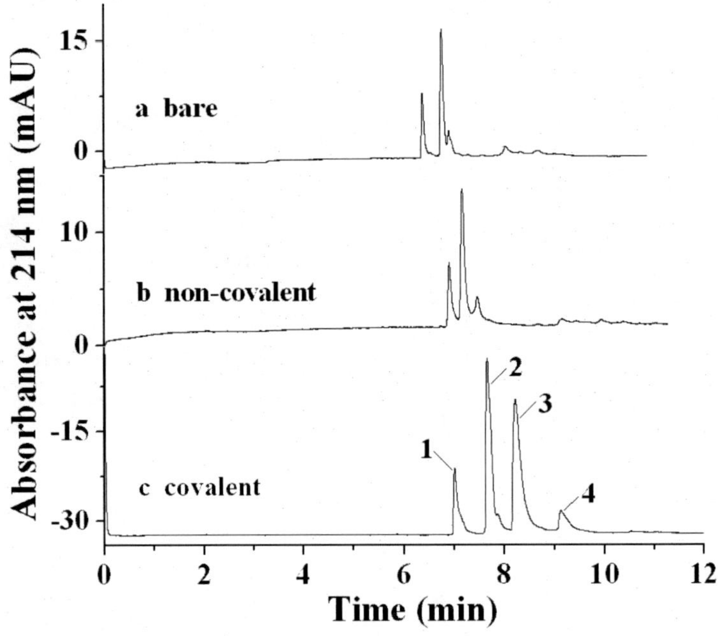

Figure 7. Separation of proteins using the bare capillary (a), PEG non-covalently coated capillary (b) and 2-layer PEG covalently coated capillary (c). Separation conditions: buffer, 40 mM phosphate (pH=3.0); injection, 20s with a height difference of 20cm; applied voltage =18 kV; UV detection, 214 nm; sample, 0.5 mg/mL for each protein; capillary, 75 μm ID× 50cm (41 cm effective); capillary temperature, 25°C. Peak identification: (1) Cyt-c; (2) Lys; (3) BSA; (4) RNase A [84].

The covalently bonded coatings suppressed protein adsorption on the inner surface of the capillary, and thus a baseline separation of lysozyme(Lys), cytochrome c(Cyt-c), bovine serum albumin(BSA) and ribonuclease A(RNase A) was achieved using capillary electrophoresis(CE), as shown in Figure 7. Compared with the bare capillary or non-covalently bonded DR/PEG coatings, the covalently linked DR/PEG capillary coatings not only improved the CE separation performance for proteins, but also exhibited good stability and repeatability.

Kovach et al. described the modification of sealed PDMS microchannels using oxygen plasma pretreatment and PEG grafting approach [85]. We present results regarding the testing of the coated microchannels after extended periods of aging and blood exposure. The PEG-grafted channels showed significantly reduced fibrinogen adsorption and platelet adhesion up to 28 days after application, highlighting the stability and functionality of the coating over time. The microchannel networks also displayed a significantly reduction in the coagulation response under whole blood flow. Further, pressure across coated microchannel networks took over 16 times longer to double than the uncoated controls. Marco et al. presented a PEG-modified PMMA (PEGMA) for the fabrication of monolithic microchannels that can withstand pressures up to 12 MPa and that are inherently biocompatible [86]. This has been achieved by successive direct UV-photo partial polymerization of PEGMA layers through a printed photomask, followed by a bonding step in which all layers are assembled by completion of the polymerization process, resulting in a rapid and low-cost method to produce microfluidic structures. PEG as a well-known antifouling material is often coated on surfaces to form highly solvated brushes, which exhibit excellent protein-repellent properties. However, little is known about the intrinsic PEG-protein interactions in aqueous solution, which is an important yet neglected problem. Wu et al. investigated the interactions between PEG and proteins in aqueous solution [87]. They found that PEG with optical molecular weight is more capable of interacting with proteins, which induces the conformational change of proteins through more stable binding sites and stronger interactions with long chain PEG. Enhanced PEG-protein interactions are likely due to the change of hydrophilicity to amphiphilicity of PEG with increasing molecular weight. Such knowledge may help to better understand the structure-property relationship of PEG and rationally design effective PEG polymers with proper molecular weight for desirable properties. Röttgermann et al. [88] had correlated the amount of FN adsorption into different PEG layers with cell adhesion and motility. PEG grafted to a hydrophobic polypropylene oxide

(PPO) anchor suppresses cell adhesion and migration almost completely and showed no fibronectin adsorption. The combination of partial adsorption of cell adhesive proteins and the cell repellent properties of PEG are suited to elevate cell motility. Thus, artificially designed surfaces of PEG block copolymers and proteins could fine-tune cell motility properties. Samanta et al. [89] investigated the thermal stability of the globular transport protein human serum albumin (HAS), in the presence of two small chains PEG200 and PEG400. The addition of PEG moderately increases the α-helical content of the protein without abruptly changing its tertiary structure. The hydration structure at the protein surface experiences a notable change at 30% PEG concentration. They concluded that both the indirect excluded volume principle as well as interaction of the polymer at the protein surface is responsible for the observed change of the unfolding process.

3.3. Application in Cell Culture

Productivity has become a focal issue in the purification of monoclonal antibodies for cell cultural applications. Protein A affinity chromatography and other column-based capture methods have come to be regarded as a bottleneck. The performance of the modified magnetic particles (MPs) for antibody purification was investigated by Dhadge and co-workers with an aqueous two-phase system composed of PEG and dextran [90]. In direct magnetic capturing, and using pure protein solutions of human immunoglobulin G (hIgG) and BSA, MP-extracellular polysaccharide bound 120 mg·hIgG·g^{-1}·MPs, whereas with BSA only 10±2 mg·BSA·g^{-1}·MPs was achieved. In multiple extraction steps, the MPs bound 92% of loaded hIgG with a final purity level of 98.5%. FucoPol coating allowed both electrostatic and hydrophobic interactions with the antibody contributing to enhance the specificity for targeted products. PEGylated adrenomedullin (AM) stimulated cAMP production, an intracellular second messenger of AM, in cultured human embryonic kidney cells expressing a specific AM receptor in a dose-dependent manner. Kubo et al. [91] found that intravenous bolus injection of 10 nmol/kg PEGylated AM lowered blood pressure in anesthetized rates; plasma half-life of PEGylated AM was significantly longer than native AM both in the first and second phases in rats. In a word, N-terminal PEGylated AM stimulated cAMP production in vitro, showing lessened acute hypotensive action and a prolonged plasma half-life in comparison with native AM peptide in vivo. Gagnon et al. [92] achieved exceptionally high capacity capture of monoclonal IgG by adding 200 nm starch-coated MPs as nucleation centers, adding PEG, and then collecting the particle-associated antibody in a magnetic field. The accretion of IgG begins on particle surfaces then continues with fusion of particle-centric accretions up to about 1 mm in a process that closely parallels PEG precipitation. Aggregate content was reduced from 4.9% to 3.6% at the nanoparticle step, then to less than 0.05% at the multimodal step. The multimodal step also removed residual PEG; overall IgG recovery was 68%. Bouzas et al. [93] investigated the impact of PEG-coated NPs on freshly purified primary cultures of rat alveolar type II (ATII) cells. The result showed that PEG-coated NPs are very modestly internalized by ATII cells and it neither leads to detectable morphological changes nor to decreased cell viability nor to alterations in basic functional parameters such as pulmonary surfactant secretion, even on exposure to high gold concentration (~0.2 mM) during relatively long periods of time (24~48 h). A photocrosslinkable PEG-laminin 111 hydrogel was developed by Francisco et al. [94]. When primary immature porcine NP cells were seeded onto PEG-laminin 111 hydrogels of varying stiffnesses, laminin111-presenting hydrogels were found to promote cell clustering and increased levels of sGAG production as compared to stiffer laminin111-presenting and PEG-only gels. The soft, laminin 111-functionalized hydrogels may promote or maintain the expression of specific markers characteristic of an immature NP cell phenotype. Age-related macular degeneration (AMD) is a leading cause of irreversible blindness. Lyzogubov's study [95] was done to characterize dry AMD-like changes in mouse retinal pigment epithelium (RPE) and retina after PEG treatment. In all other experiments 0.5 mg PEG was injected and animals were sacrificed on days 1, 3, 5 or 14. Subretinal injection of 0.5 mg PEG induced a 32% reduction of outer nuclear layer (ONL) thickness, 61% decrease of photoreceptor outer and inner segment length, 49% decrease of nuclear density in the ONL and 31% increase of RPE cell density by day 5 after injection. Degeneration of RPE cells was found in PEG injected eyes. PEG leads to morphological and gene expression changes in RPE and retina consistent with dry AMD. Zinc is one of the trace elements contained in bone apatite, with anti-inflammatory and stimulatory effects on bone forming. Zinc containing bioactive glasses were prepared by means of sol-gel procedure, in the presence of PEG considered as surfactant agent by Veres and co-workers [96]. The results proved that gamma irradiation induces structural changes within the glass matrix, which are strongly dependent on the sample composition. ZnO addition to the glasses network was found to confer structural stability against gamma irradiation, while the use of PEG during the samples preparation favored the generation of network defects.

3.4. Application in Biomedical Engineering

PEG has average molecular weights within the range of 200-35000. The chemistry and properties of these polymeric networks can be tuned using longer chains of functionality cross-linkers which make this material suitable for biomedical applications. NPs that target bone tissue were developed using PEG and bone-targeting moieties based on aspartic acid by Fu and co-workers [97]. To evaluate the in vitro affinity of hydroxyapatite powder and its ability to be endocytosed by D1 bone marrow stem cell line, a bone targeting in vivo feasibility assessment was performed using fluorescent imaging experiments in zebrafish and Sprague-Dawley rats. The average particle size of NPs was < 200 nm and the dendritic aspartic acid moiety of the modified NPs exhibited the best apatite mineral binding ability. PEG- coated iron-oxide super-paramagnetic NPs were synthesized and its cytotoxicity and biodistribution assessed by Mojica Pisciotti and co-workers [98]. PEG- SPIONs has specific power absorption of 320 W/g and presents high cell viability for concentration as high as 200 µg/mL. Myocardial matrix-PEG hybrids were synthesized by two different methods, cross-linking proteins with an amine-reactive PEG-star and photo-induced radical polymerization of two different multi-armed PEG-acrylates by Grover et al. [99] Their work demonstrates that PEG incorporation into extracellular matrix based hydrogels can expand material properties, thereby opening up new possibilities for in vitro and in vivo applications. The use of PEG functionalized NPs has gained much attention due to promising biodistribution and stealth properties in vivo [100]. PEG coated NPs have been designed to enhance stealth properties or drug release behavior or other biomedical engineering applications. A novel theranostic agent based on Prussian blue nanocubes (PB NCs) for in vivo bimodal imaging-guided photothermal therapy by Cheng and co-workers [101]. After being coated with PEG, the obtained PB-PEG NCs are highly stable in different physiological solutions. Comprehensive blood tests and careful histological examinations reveal no apparent toxicity of PB-PEG NCs to mice at their tested dose within two months. It is quite a promising multifunctional nanoprobe for imaging-guided cancer therapy. Jue et al. [102] developed a rapid point-of-need detection assay as an extension to proof- of-concept studies which used a micellar aqueous two-phase system. The gold nanoprobes were modified by PEG to provide steric stabilization, and used to demonstrate a 10-fold improvement in the lateral-flow immunoassay detection limit.

There still exist other applications of biocompatible PEG according to our recent works. PEG hydrogel is employed as the substrate material in our study about the effect of patterned non-adhesive hydrogel nanosurface on the self-assembly of highly ordered colloids [103].

By simply controlling the colloidal concentration of the nanoassembly solutions and the dimensions of the wells, a range of highly organized nanocolloidal patterns are formed inside the PEG wells. Benefiting from the unique non-adhesive surface property, the pattern-assisted nanoassembly method enables a highly predictable and robust process for colloidal nanofabrication, and the obtained nanocolloidal arrays with well-organized patterns could potentially find applications in biological sensing. Aminosilane-modified PG and other PEG block prepolymers are synthesized as new membrane materials for gas separation [104].

The silane-modified prepolymers form flexible nanocomposite membranes with a maximum CO_2 permeability of 274 barrer and CO_2/N_2 selectivity of about 30. Subsequently, Pebax-1657 nanocomposite membranes incorporated with silica nanoparticles, PS colloids, and CNTs were fabricated successfully for CO_2/N_2 and CO_2/H_2 separation [105]. At 5 wt% SWNTs and MWNTs, the CO_2 permeability of the membrane also increased by 28 and 13%, respectively.

CONCLUSION

In this chapter, we give a systematic, balanced and comprehensive summary of the main aspects of photosensitive PEG related to its preparation and applications in micro-nanofabrication. The use of photosensitive polymeric materials possessing ease of fabrication and modification for functionalization has gained a special interest for micro-nanofabrication applications.

The synthesis of novel photosensitive polymers and copolymers is demonstrated due to its defined architectures and improved properties. Many practical micro-nano materials and biochips based on photosensitive PEG with potential applications in drug delivery, protein separation, cell culture and biomedical engineering have been generated.

ACKNOWLEDGMENTS

This work is financially supported by the National Key Basic Research Development Program of China (973 special preliminary study plan, 2012CB722705), the Natural Science Foundation of China (21375069, 21404065), the Fok Ying Tong Education Foundation of Ministry of Education of China (131045), the Natural Science Foundation for Distinguished Young Scientists of Shandong Province (JQ201403), the Graduate Education Innovation Project of Shandong Province (SDYY14028), the Scientific Research Foundation for the Returned Overseas Chinese Scholars of State Education Ministry (20111568), Science and Technology Program of Qingdao (1314159jch), the China Postdoctoral Science Foundation (2014M561886) and the Doctoral Scientific Research Foundation of Qingdao.

REFERENCES

[1]   Harris J M, *Macromol J. Sci.-Rev. Macromol. Chem. Phys.*, C25, 1985, 325–373.

[2]   Albertsson P A, "Partition of Cell Particles and Macromolecules," Wiley Interscience, New York, 3rd ed., 1986.

[3]   Walter H, Brooks D E, and Fisher D (Editors), "Partitioning in Aqueous Two-Phase Systems," Academic Press, Orlando, FL, 1985.

[4]   Fisher D and Sutherland I A (Editors), "Separations Using Aqueous Phase Systems: Applications in Cell Biology and Biotechnology," Plenum, London, 1989.

[5]   GrahamN B in "Hydrogels in Medicine and Pharmacy," Vol. II, N. A. Pappas, ed., CRC Press; Boca Raton, FL, 1988, p. 95.

[6]   Zalipsky S, Alericia F, Barany G, in "Peptides: Structure and Function," 9th American Peptide Symposium, C. M. Deber, V. J. Hruby and K. D. Kopple, 1985, p. 257.

[7]   Yoshinaga Y, Harris J M. Effects of coupling chemistry on activity of a polyethylene glycol-modified enzyme. *J. Bioact. Comp. Polym.*, 1989, 4: 17–24.

[8]   Jacobs H A, Okano T, Kim S W. Antithrombogenic surfaces: Characterization and bioactivity of surface immobilized PGE1-heparin conjugate. *Journal of biomedical materials research*, 1989, 23(6): 611–630.

[9]   Stark M B and Holmberg J K, Biotech. Bioeng. 1989, 34, 942.

[10] Zalipsky S, Gilon C, Zilkha A. Attachment of drugs to polyethylene glycols. *European polymer journal*, 1983, 19(12): 1177–1183.

[11] Macewen E G, Rosenthal R, Matus R, et al. A preliminary study on the evaluation of asparaginase. Polyethylene glycol conjugate against canine malignant lymphoma. *Cancer,* 1987, 59(12): 2011–2015.

[12] Suzuki T, Kanbara N, Tomono T, et al. Physicochemical and biological properties of poly (ethylene glycol)-coupled immunoglobuling G. Biochimica et Biophysica Acta (BBA)-Protein Structure and Molecular *Enzymology,* 1984, 788(2): 248–255.

[13] Andrade J D, Nagaoka S, Cooper S, et al. Surfaces and Blood Compatibility Current Hypotheses. ASAIO Journal, 1987, 33(2): 75.

[14] KatreN V, Knauf M J, andLaird W *J, Proc. Natl. Acad. Sci. USA,* 1987, 84, 1487.

[15] Gölander C G, Kiss E. Protein adsorption on functionalized and ESCA-characterized polymer films studied by ellipsometry. *Journal of colloid and interface science*, 1988, 121(1): 240–253.

[16] Merrill E W, Salzman E W, J. Am. Soc. Artif. Intern. Organs, 1983, 6, 60.

[17] Gombotz W R, Guanghui W, Hoffman A S. Immobilization of poly (ethylene oxide) on poly (ethylene terephthalate) using a plasma polymerization process. *Journal of applied polymer science*, 1989, 37(1): 91–107.

[18] Herren B J, Shafer S G, van Alstine J, et al. Control of electroosmosis in coated quartz capillaries. *Journal of colloid and interface science*, 1987, 115(1): 46–55.

[19] Harris J M, Brooks D E, Boyce J F, Snyder R S, and Van Alstine J M, in "Dynamic Aspects of Polymer Surfaces," J. D. Andrade, Ed., Plenum, 1988, pp 111–119.

[20] Harris J M, Dust J M, McGill R A, et al. *New polyethylene glycols for biomedical applications*. 1991, 467: 418.

[21] Hamidi, M, Azadi, A, and Rafiei, P. Pharmacokinetic consequences of pegylation. *Drug Delivery*. 2006, 13, 399–409.

[22] Ryan S M, Mantovani G, Wang X, Haddleton D M, and Brayden D J. Advances in PEGylation of important biotech molecules: Delivery aspects. *Expert Opin. Drug Delivery*. 2008, 5, 371–383.

[23] Veronese F M, and Mero A. The impact of PEGylation on biological therapies. *BioDrugs,* 2008, 22, 315–329.

[24] *Poly(ethylene glycol) chemistry: Biotechnical and biomedical applications* (Harris, J. M., Ed.) (1992) pp 1-408, Springer, New York.

[25] Greenwald R B, Choe Y H, McGuire J, and Conover C D. Effective drug delivery by PEGylated drug conjugates. *Adv. Drug Delivery Rev.* 2003, 55, 217–250.

[26] Jain A, and Jain S K. PEGylation: An approach for drug delivery. A review. *Crit. ReV. Ther. Drug Carrier Syst.* 2008, 25, 403–447.

[27] Georgianna W E, Lusic H, McIver A L, et al. Photocleavable polyethylene glycol for the light-regulation of protein function. *Bioconjugate chemistry*, 2010, 21(8): 1404–1407.

[28] Pathak C P. et al. *J. Am. Chem. Soc.* 1992, *114*, 8311–8312.

[29] Desai N P, Hubbell J A. Solution technique to incorporate polyethylene oxide and other water-soluble polymers into surfaces of polymeric biomaterials. *Biomaterials,* 1991, 12(2): 144–153.

[30] Matsuda T, Nakayama Y. *J. Polym. Sci.: Part A* 1992, *30*, 2451–2457.

[31] Matsuda T, Moghddam M J. *Trans. Am. Soc. Artif. Intern Organs* 1991, *37*, M437–M438.

[32] Sawhney A S, Pathak C P, Hubbell J A. Bioerodible hydrogels based on photopolymerized poly (ethylene glycol)-co-poly (alpha.-hydroxy acid) diacrylate macromers. *Macromolecules,* 1993, 26(4): 581–587.

[33] Coury A J. IBC Conference: Biocompatibility of Blood-Interacting Materials. *In situ* Formed Bioabsorbable Hydrogel Lining of Damaged Arteries: Potential and Preclinical Progress; Boston, MA, U.S.A., 1995, June 26–27.

[34] Andreopoulos F M, Deible C R, Stauffer M T, et al. Photoscissable Hydrogel Synthesis via Rapid Photopolymerization of Novel PEG-Based Polymers in the Absence of Photoinitiators. *Journal of the American Chemical Society,* 1996, 118(26): 6235–6240.

[35] Doytcheva M, Dotcheva D, Stamenova R, et al. Ultraviolet-induced crosslinking of solid poly (ethylene oxide). *Journal of applied polymer science,* 1997, 64(12): 2299–2307.

[36] Sloop S L, *J. Appl. Polym. Sci.* 1994, 53, 1563.

[37] Emami S H, Salovey R, Hogen-Esch T E. Degradable poly (ethylene oxide) hydrogels formed by crosslinking with tert-butylperoxybenzoate. *Journal of Polymer Science Part A: Polymer Chemistry*, 2003, 41(4): 520–527.

[38] Lin-Gibson S, Bencherif S, Cooper J A, et al. Synthesis and characterization of PEG dimethacrylates and their hydrogels. *Biomacromolecules,* 2004, 5(4): 1280–1287.

[39] Sawhney A S, Pathak C P, Hubbell J A. Bioerodible hydrogels based on photopolymerized poly (ethylene glycol)-co-poly (alpha.-hydroxy acid) diacrylate macromers. *Macromolecules,* 1993, 26(4): 581–587.

[40] Andreopoulos F M, Deible C R, Stauffer M T, et al. Photoscissable Hydrogel Synthesis via Rapid Photopolymerization of Novel PEG-Based Polymers in the Absence of Photoinitiators. *Journal of the American Chemical Society,* 1996, 118(26): 6235–6240.

[41] Andreopoulos F M, Beckman E J, Russell A J. Photoswitchable PEG - CA hydrogels and factors that affect their photosensitivity. *Journal of Polymer Science Part A: Polymer Chemistry,* 2000, 38(9): 1466–1476.

[42] Zheng Y, Micic M, Mello S V, et al. PEG-based hydrogel synthesis via the photodimerization of anthracene groups. *Macromolecules,* 2002, 35(13): 5228–5234.

[43] Nagata M, Yamamoto Y. Photoreversible poly (ethylene glycol) s with pendent coumarin group and their hydrogels. *Reactive and Functional Polymers,* 2008, 68(5): 915–921.

[44] Revzin A, Russell R J, Yadavalli V K, et al. Fabrication of poly (ethylene glycol) hydrogel microstructures using photolithography. *Langmuir,* 2001, 17(18): 5440–5447.

[45] Ward J H, Bashir R, Peppas N A. *J. Biomed. Mater. Res.* 2001, 56, 351–360.

[46] Koh W G, Revzin A, Simonian A, Reeves T, Pishko M. Biomed. *Microdevices* 2003, 5, 11–19.

[47] Revzin A, Tompkins R G, Toner M. Surface engineering with poly (ethylene glycol) photolithography to create high-density cell arrays on glass. Langmuir, 2003, 19(23): 9855–9862.

[48] Håfors B, Rowell R M, Barbour R J. The role of the Wasa in the development of the polyethylene glycol preservation method. *Advances in Chemistry Series,* 1990 (225): 195–216.

[49] Filpula D, Zhao H. Releasable PEGylation of proteins with customized linkers. *Advanced drug delivery reviews,* 2008, 60(1): 29–49.

[50] Digilio G, Barbero L, Bracco C, et al. NMR structure of two novel polyethylene glycol conjugates of the human growth hormone-releasing factor, hGRF (1-29)-NH_2. *Journal of the American Chemical Society,* 2003, 125(12): 3458–3470.

[51] Gonnelli M, Strambini G B. No effect of covalently linked poly (ethylene glycol) chains on protein internal dynamics. Biochimica et

Biophysica Acta (BBA)-Proteins and Proteomics, 2009, 1794(3): 569–576.

[52] Lin C C, Anseth K S. PEG hydrogels for the controlled release of biomolecules in regenerative medicine. *Pharmaceutical research*, 2009, 26(3): 631–643.

[53] Lawrence D S. The preparation and in vivo applications of caged peptides and proteins. *Current opinion in chemical biology*, 2005, 9(6): 570–575.

[54] Mayer G, Heckel A. Biologically active molecules with a "light switch." *Angewandte Chemie International Edition*, 2006, 45(30): 4900–4921.

[55] Tang X J, Dmochowski I J. Regulating gene expression with light-activated oligonucleotides. Molecular BioSystems, 2007, 3(2): 100–110.

[56] Casey J P, Blidner R A, Monroe W T. Caged siRNAs for spatiotemporal control of gene silencing. *Molecular pharmaceutics*, 2009, 6(3): 669–685.

[57] Lee H M, Larson D R, Lawrence D S. Illuminating the chemistry of life: design, synthesis, and applications of "caged" and related photoresponsive compounds. *ACS chemical biology*, 2009, 4(6): 409–427.

[58] Deiters A. Light activation as a method of regulating and studying gene expression. *Curr. Opin. Chem. Biol.* 2009, 13, 678–686.

[59] Deiters A. Principles and applications of the photochemical control of cellular processes. *Chem. Bio. Chem.* 2010, 11, 47–53.

[60] Young D D, and Deiters A. Photochemical control of biological processes. Org. Biomol. Chem. 2007, 5, 999–1005.

[61] Fischer B, Perry B, Phillips G, Sumner I, and Goodenough P. Physiological consequence of expression of soluble and active hen egg white lysozyme in escherichia coli. *Appl. Microbiol. Biotechnol.* 1993, 39, 537–540.

[62] Sava G. Pharmacological aspects and therapeutic applications of lysozymes. *EXS* 1996, 75, 433–449.

[63] Van Dael H. Chimeras of human lysozyme and alphalactalbumin: An interesting tool for studying partially folded states during protein folding. *Cell. Mol. Life Sci.* 1998, 54, 1217–1230.

[64] Masschalck B, and Michiels C W. Antimicrobial properties of lysozyme in relation to foodborne vegetative bacteria. *Crit. ReV. Microbiol.* 2003, 29, 191–214.

[65] Wang J, Dauter M, Alkire R, Joachimiak A, and Dauter Z. Triclinic lysozyme at 0.65 A resolution. *Acta Crystallogr. Sect. D: Biol. Crystallogr.* 2007, 63, 1254–1268.

[66] Shugar D. The measurement of lysozyme activity and the ultra-violet inactivation of lysozyme. *Biochim. Biophys. Acta* 1952, 8, 302–309.

[67] Roberts M J., and Harris, J. M. Attachment of degradable poly(ethylene glycol) to proteins has the potential to increase therapeutic efficacy. *J. Pharm. Sci.* 1998, 87, 1440–1445.

[68] Tanaka H, Tsuda M. Photosensitive composition comprising light-sensitive polymer: U.*S. Patent* 3,767,415[P]. 1973-10-23.

[69] Tanaka H, Honda K *J. Polym. Sci.* 1977, 15, 2685–2689.

[70] Tanaka H, Tsuda M, Nakanishi H. Photochemistry of poly (vinyl cinnamylideneacetate) and related compounds. Journal of Polymer Science Part A-1: *Polymer Chemistry*, 1972, 10(6): 1729–1743.

[71] Tanaka H, Sato Y. Photosensitivity of polyvinylesters of substituted cinnamylideneacetic acids. *Journal of Polymer Science Part A-1: Polymer Chemistry,* 1972, 10(11): 3279–3287.

[72] Tanaka H. Otomegawa, E. *J. Polym. Sci.* 1974, *12*, 1125–1130.

[73] Reiser A. Photoreactive Polymers. The Science and Technology of Resists: Negative Photoresists; John Wiley & Sons: New York, 1989; pp 24–28.

[74] Ren Y M, Yu B, Cong H L, et al. Preparation of monodisperse PEG microspheres by a T-junction microfluidic chip. *Advanced Materials Research,* 2012, 465: 178–181.

[75] Yu B, Cong H, Liu X, et al. Preparation of monodisperse PEG hydrogel composite microspheres via microfluidic chip with rounded channels. *Journal of Micromechanics and Microengineering,* 2013, 23(9): 095016.

[76] Ma Y, Gill H S. Coating solid dispersions on microneedles via a molten dip-coating Method: development and in vitro evaluation for transdermal delivery of a water-insoluble drug. *Journal of pharmaceutical sciences,* 2014, 103(11): 3621–3630.

[77] Dorniani D, Kura A U, Hussein M Z B, et al. Controlled-release formulation of perindopril erbumine loaded PEG-coated magnetite nanoparticles for biomedical applications. *Journal of Materials Science,* 2014, 49(24): 8487–8497.

[78] Pacelli S, Paolicelli P, Pepi F, et al. Gellan gum and polyethylene glycol dimethacrylate double network hydrogels with improved mechanical properties. *Journal of Polymer Research*, 2014, 21(5): 1–13.

[79] Cui Z K, Edwards K, Orellana A N, et al. Impact of interfacial cholesterol-anchored polyethylene glycol on sterol-rich non-phospholipid liposomes[J]. *Journal of Colloid and Interface Science*, 2014, 428: 111–120.

[80] Tamilvanan S, Kumar V, Sharma D, et al. In vitro evaluation of polyethylene glycol based microparticles containing azithromycin. *Drug Delivery and Translational Research*, 2014, 4(2): 139–148.

[81] Schou-Pedersen A M V, Hansen S H, Moesgaard B, et al. Kinetics of the esterification of active pharmaceutical ingredients containing carboxylic Acid functionality in polyethylene glycol: formulation implications. *Journal of pharmaceutical sciences*, 2014, 103(8): 2424–2433.

[82] Moorkoth D, Nampoothiri K M. Synthesis, Colloidal Properties and Cytotoxicity of Biopolymer Nanoparticles. *Applied biochemistry and biotechnology*, 2014, 174(6): 2181–2194.

[83] Hooftman G, Herman S, Schacht E. Review: poly (ethylene glycol)s with reactive endgroups. II. Practical consideration for the preparation of protein-PEG conjugates. *Journal of bioactive and compatible polymers*, 1996, 11(2): 135–159.

[84] Yu B, Cui W, Cong H, et al. A novel diazoresin/polyethylene glycol covalent capillary coating for analysis of proteins by capillary electrophoresis. *RSC Advances*, 2013, 3(43): 20010–20015.

[85] Kovach K M, Capadona J R, Sen Gupta A, et al. The effects of PEG-based surface modification of PDMS microchannels on long-term hemocompatibility. *Journal of Biomedical Materials Research Part A*, 2014, 102(12): 4195–4205.

[86] De Marco C, Credi C, Briatico-Vangosa F, et al. Fabrication of biocompatible monolithic microchannels with high pressure-resistance using direct polymerization of PEG-modified PMMA. *Journal of Applied Polymer Science*, 2014, 131(21), DOI: 10.1002/app.41031.

[87] Wu J, Zhao C, Lin W, et al. Binding characteristics between polyethylene glycol (PEG) and proteins in aqueous solution. *Journal of Materials Chemistry B*, 2014, 2(20): 2983–2992.

[88] Röttgermann P J F, Hertrich S, Berts I, et al. Cell Motility on Polyethylene Glycol Block Copolymers Correlates to Fibronectin Surface Adsorption. *Macromolecular bioscience*, 2014, 14(12): 1755–1763.

[89] Samanta N, Mahanta D D, Hazra S, et al. Short Chain Polyethylene Glycols Unusually Assist Thermal Unfolding of Human Serum Albumin. *Biochimie*, 2014, 104: 81–89.

[90] Dhadge V L, Morgado P I, Freitas F, et al. An extracellular polymer at the interface of magnetic bioseparations. *Journal of the Royal Society Interface*, 2014, 11(100): 20140743.

[91] Kubo K, Tokashiki M, Kuwasako K, et al. Biological properties of adrenomedullin conjugated with polyethylene glycol. *Peptides,* 2014, 57: 118–121.

[92] Gagnon P, Toh P, Lee J. High productivity purification of immunoglobulin G monoclonal antibodies on starch-coated magnetic nanoparticles by steric exclusion of polyethylene glycol. *Journal of Chromatography A,* 2014, 1324: 171–180.

[93] Bouzas V, Haller T, Hobi N, et al. Nontoxic impact of PEG-coated gold nanospheres on functional pulmonary surfactant-secreting alveolar type II cells. *Nanotoxicology,* 2014, 8(8): 813–823.

[94] Francisco A T, Hwang P Y, Jeong C G, et al. Photocrosslinkable laminin-functionalized polyethylene glycol hydrogel for intervertebral disc regeneration. *Acta biomaterialia,* 2014, 10(3): 1102–1111.

[95] Lyzogubov V V, Bora N S, Tytarenko R G, et al. Polyethylene glycol induced mouse model of retinal degeneration. *Experimental eye research,* 2014, 127: 143–152.

[96] Veres R, Vanea E, Gruian C, et al. The effects of PEG assisted synthesis and zinc addition on gamma irradiated bioactive glasses. *Composites Part B: Engineering,* 2014, 66: 83–88.

[97] Fu Y C, Fu T F, Wang H J, et al. Aspartic acid-based modified PLGA–PEG nanoparticles for bone targeting: In vitro and in vivo evaluation. *Acta biomaterialia,* 2014, 10(11): 4583–4596.

[98] Mojica Pisciotti M L, Lima E, Vasquez Mansilla M, et al. In vitro and in vivo experiments with iron oxide nanoparticles functionalized with DEXTRAN or polyethylene glycol for medical applications: Magnetic targeting. *Journal of Biomedical Materials Research Part B: Applied Biomaterials,* 2014, 102(4): 860–868.

[99] Grover G N, Rao N, Christman K L. Myocardial matrix–polyethylene glycol hybrid hydrogels for tissue engineering. *Nanotechnology,* 2014, 25(1): 014011.

[100] Cruje C, Chithrani D B. Polyethylene Glycol Functionalized Nanoparticles for Improved Cancer Treatment. *Reviews in Nanoscience and Nanotechnology,* 2014, 3(1): 20–30.

[101] Cheng L, Gong H, Zhu W, et al. PEGylated Prussian blue nanocubes as a theranostic agent for simultaneous cancer imaging and photothermal therapy. *Biomaterials,* 2014, 35(37): 9844–9852.

[102] Jue E, Yamanishi C D, Chiu R Y T, et al. Using an aqueous two‐phase polymer-salt system to rapidly concentrate viruses for improving the detection limit of the lateral-flow immunoassay. *Biotechnology and bioengineering*, 2014, 111(12): 2499–2507.

[103] Cong H, Revzin A, Pan T. Non-adhesive PEG hydrogel nanostructures for self-assembly of highly ordered colloids. *Nanotechnology*, 2009, 20(7): 075307.

[104] Cong H, Yu B. Aminosilane cross-linked PEG/PEPEG/PPEPG membranes for CO_2/N_2 and CO_2/H_2 separation. *Industrial & Engineering Chemistry Research*, 2010, 49(19): 9363–9369.

[105] Yu B, Cong H, Li Z, et al. Pebax-1657 nanocomposite membranes incorporated with nanoparticles/colloids/carbon nanotubes for CO_2/N_2 and CO_2/H_2 separation. *Journal of Applied Polymer Science*, 2013, 130(4): 2867–2876.

In: Polyethers and Polyethylene Glycol
Editor: Joyce Williams

ISBN: 978-1-63483-392-9
© 2015 Nova Science Publishers, Inc.

Chapter 2

BIODEGRADATION OF POLYETHYLENE GLYCOLS AND POLYETHOXYLATED SURFACTANTS

F. Kawai

Center for Fiber and Textile Science, Kyoto Institute of Technology,
Matsugasaki, Sakyo-ku, Kyoto, Japan

ABSTRACT

Research on biodegradation of polyethylene glycols (PEGs) and polyethoxylated surfactants has a long history of over 50 years. Aerobic and anaerobic microbial degradation of PEGss and polyethoxylated surfactants has been documented, although the mechanism of anaerobic degradation has not been fully elucidated. It is noteworthy that ether bonds are resistant for hydrolysis except acidic conditions, which are not the cases for most of expected biodegradation. PEG and polyethoxy chains are depolymerized step by step as one glycol unit from the hydroxyl terminals under either aerobic or anaerobic conditions. The aerobic metabolism of PEGs and polyethoxylated compounds is started by PEG dehydrogenase (PEG-DH). The *peg* operon was comprised of six genes including *pegA* (the gene encoding PEG-DH) in Sphingomonads. As *pegA* is likely to have been distributed among different PEG-degrading bacterial species, the *peg* operon or genes relevant to the metabolism of PEG are probably existent in a variety of bacteria. On the other hand, wood-rot fungi catalyze non-metabolic depolymerization of PEG *via* peroxidase or Fenton reaction, which happens at random on the polyethoxy chain. Thus biodegradation of PEGs and polyethoxylated compounds is no more of a big problem. PEGs appear to be metabolically inert and nontoxic, but some report described that `PEGs were sulfated *in vitro* by the rat and guinea pig livers, and carboxylated *in vivo* in rabbits and burn patients. In addition, as some depolymerized metabolites were reported as toxic and less biodegradable than original compounds, pursuit of metabolites of PEGs and polyethoxylates either in the environment or *in vivo* is meaningful.

I. INTRODUCTION [1]

The most unique point of polyethylene glycol (PEG) is its variety of molecular sizes from oligomers to high molecular masses up to ten million. Accordingly the physical properties of PEGs vary from viscous liquids to solids based on their molecular sizes, although every PEG from oligomers up to polymers with a molecular mass of ten million is completely water-soluble. Polyethylene oxide (PEO) and polyoxyethylene (POE) are chemically synonymous to PEG, but historically the name PEG is often referred to oligomers and polymers with a molecular mass below 20,000. Due to the hydrophilic property of PEGs, the majority of PEGs produced are used to make nonionic surfactants combined with hydrophobic groups such as alkylphenols, aliphatic alcohols, sorbitan acyl esters etc., the very important groups of industrial chemicals with applications from domestic detergents to agrochemicals, food emulsifiers and other industrial preparations. PEGs are also copolymerized with another polyether, polypropylene glycol (PPG) and used as detergents. Commercially available PPGs can be divided into two groups, the diol and triol types, based on the straight or branched chain structure of the polymer. The water solubility of PPGs is lost when the molecular mass is increased to more than approximately 700 (triol type) and 1,000 (diol type) due to the inclusion of a methyl group in each monomer unit. Therefore, copolymers of PEG and PPG are used as detergents, where PEG is a hydrophilic constituent and PPG a hydrophobic one. Another copolymer is also used as a water-soluble flame-resisting pressure liquid, where ethylene oxide and propylene oxide are randomly copolymerized. These products should ultimately constitute a significant burden on domestic and industrial wastewater systems. Therefore, their biodegradability characteristics in sewage treatment plants and aquatic environments have been observed over the past 50 years. Because of low toxicity and skin irritation, PEGs are widely used in the pharmaceutical industry in the preparation of ointments, suppositories, tablet excipients, laxatives and solvents for injection, and also for the preparation of cosmetics such as creams, lotions, powders, cakes, and lipsticks. They are also used as intermediates in the production of resins such as alkyd resin and polyurethane resin, and as components in the manufacture of lubricants, antifreeze agents, wetting agents, printing inks, adhesives, shoe polish, softening agents, sizing agents, and plasticizers. As PEG replaces water in wooden objectives, making the wood dimensionally stable and preventing warping or shrinking of the wood when it dries, PEG has been used to preserve relics that have been salvaged from underwater or dug from the underground. In addition, PEG is used as a stabilizer for green wood, preventing shrinkage. Furthermore, this material has been used in making resin gels to immobilize enzymes or microbial cells and in the chemical modification of enzymes and therapeutic proteins. The coupling of a PEG to another larger molecule is called as PEGylation, for example, PEGylated interferons are commonly used injectable treatments for infection. Efficiency of PEG has been proposed in the repair of the spinal injuries (SCRIPPS NEWS HOWARD SERVICE, published 10:00 pm, December 3, 2004) and in the prevention of cancer (https://www.ncbi.nlm.nih.gov/ubmed/10866305). Although PEGs appear to be metabolically inert and nontoxic, they are sulfated *in vitro* by the rat and guinea pig liver [2], and repeated topical application of a PEG-based antimicrobial cream to open wounds in rabbits and burn patients has been found to cause a syndrome related to the metabolism of PEGs to various compounds, including mono- and diacids [3]. Furthermore, the possibility that PEG 200 (approximately pentamer size) and PEG oligomers are toxic has also been suggested [4]. Biodegradation of longer PEG might pose an additional risk to the environment due to metabolite production. On the other hand, alkylphenol mono- or diethoxylate and their carboxylated forms were detected as the end products of the parent detergents, as described below. Thus free forms of PEGs and polyethoxylated detergents have been most important targets to know their fate in the environment.

II. MECHANISMS OF PEG BIODEGRADATION

As water-soluble PEGs finally enter in an aquatic environment, microbial degradation of PEGs is most important in the ecosystem. In principle, the properties of PEGs impose large constraints on enzymatic degradation. Since it is a large molecule, as required by most applications, the membranes of the organism form a barrier against substrate uptake, except in cases in which extracellular enzymes are used for degradation, and the size of the active site of the enzyme attacking the polymers must be rather large [5]. PEG has a random coil formation.

PEG includes average two-three water molecules bound per ether group when dissolved in water, making a bigger molecule (more than twice molecular mass of PEG) and occupying a bigger space than the original PEG [6, 7].

Recent research on polymer biodegradation has revealed that big molecules such as polyvinyl alcohol (PVA: approximately 75,000 Da) or PEG 20,000 are taken up by microbial cells and metabolized within cells (most probably in the periplasmic space) [8].

As the ether bond cannot be hydrolyzed under neutral conditions where most of PEGs and their products are used, the alcohol groups at the termini can be the metabolically active sites, but they are randomly distributed in the space taken up by the macromolecule-water complex, making it difficult for the enzyme to attack the termini. Irrespective of these disadvantages, the biodegradability of PEGs with molecular masses up to 20,000 has already been well certificated with pure cultures in rather short times (in an week or so) [1].

Regarding the biodegradable molecular masses of PEG, Bernhard et al. [9] tested the removal of the dissolved organic carbon and CO_2 production from PEGs using sludge of a conventional wastewater treatment plant and filtered microorganisms from a seawater aquarium, according to standardized ISO and OECD guidelines.

Consequently PEGs up to 57,800 was degraded within 65 days by sludge microorganisms in freshwater medium and PEGs up to 14,600 were degraded in 135 days by marine microorganisms in seawater medium. This result suggests the upper limit of biodegradable PEG molecular masses, although whether the original molecular size was maintained during the long incubation or partly susceptible to the decomposition to a smaller molecule is to be confirmed. Microbes isolated from the wastewater treatment sludge efficiently decomposed PEGs up to 20,000, under either aerobic or anaerobic conditions [10]. The efficiency of the aerobic process was much higher compared with that of the anaerobic process. Taken together, PEGs up to 20,000 or so are considered sufficiently biodegradable in the aquatic environments (freshwater, marine water, wastewater plant etc.), under either aerobic or anaerobic conditions and bigger PEGs to some extent are also susceptible to biodegradation slowly in the ecosystem.

Aerobic Biodegradation of PEG

An ether bond is rather a tough bond and cannot be hydrolyzed except acidic conditions. In addition, as oxygen exists in the most area of the earth, either in the air or in the water like seas, lakes, ponds, rivers, or streams, the primary metabolism of PEG is based on aerobic degradation of PEG. Several reports indicated that aerobic degradation of PEGs is faster than anaerobic degradation of them [9, 10]. Whereas biodegradation of PEGs and ethoxy surfactantss in freshwater is in general quite fast, the biodegradation occur usually slower in saline water [9, 11]. Free forms of PEGs have two terminal hydroxyl groups and ethoxylated detergents have one terminal hydroxyl group, which are susceptible to oxidation to aldehydes and then carboxylates. Detection of shorter PEG (either free forms or ethoxylates) and their carboxylated forms [8, 11-15] indicates that PEGs are aerobically metabolizable in the ecosystem. Distribution of these metabolites in analytical samples is not always uniform, depending on many factors such as molecular size of ethoxy chains, microbial origins, medium compositions (freshwater or saline waters etc.), but the underlining mechanism is considered to be the same. In some references [11, 30], the mechanisms for production of shorter PEGs or ethoxy detergents with shorter ethoxy chains and their carboxylates were described as hydrolytic (formation of shorter PEG fragments) and oxidative-hydrolytic (formation of carboxylated forms and shortening of their ethoxy chainss), but an ether bond is in general resistant for hydrolysis. In the metabolic processes, either aerobic or anaerobic, other enzymes than hydrolases formed a shorter ethoxy chain, as described below. Therefore, hydrolysis is incorrect and depolymerization is appropriate for the formation of shorter ethoxy chain.

Various types of PEG-degraders that are able to assimilate a variety of molecular sizes have been isolated since the first report of PEG 400 by Payne [16]. Although PEGs with a molecular mass higher than 1,000 were long considered to be bioresistant, those up to 20,000 or more have since been found to be biodegradable. PEGs with a high molecular mass, from 4,000 to 20,000, are assimilated by a limited number of species: *Pseudomonas aeruginosa* (up to 20,000) [17], soil bacteria (up to 6,000) [18], *Pseudomonas stutzeri* (up to 13,500) [19], and *Sphingomonas terrae* and *S. macrogoltabida* (up to 20,000) [20]; the strains were originally identified as *Flavobacterium* species [21]. Unfortunately, no further work has been noted except Sphingomonads. Sphingomonads include sphingolipids in their outer membranes instead of the lipopolysaccharides found in most Gram-negative bacteria. Various lipophilic xenobiotic-assimilating bacteria are included in this genus [22]. Although a barrier against the macromolecules might exist for the assimilation of polyethers, the characteristic membrane structures of this genus are perhaps correlated with the uptake of large PEG by the membranes. This is partly confirmed by findings that *Rhodopseudomonas acidophila* M402 is able to oxidize PEGs up to a certain limit of size *via* its alcohol dehydrogenase, but cannot grow on these compounds [23]; in addition, cell-free extracts of PEG 400-, 1,000- and 4,000-utilizing bacteria can dehydrogenate higher PEGs (6,000 and 20,000) that cannot be utilized as sole carbon and energy sources [24]. Most recently, uptake of macromolecular PEG and PVA into the periplasm of PEG or PVA-utilizing Sphingomonads were suggested to show that the uptake of these polymers are relevant to the metabolism by these bacteria [8]. A TonB-dependent receptor-like protein (outer membrane-associated tubular protein) is involved in the *peg* operon and regulated by PEG in Sphingomonads, as described below. Uptake of ^{14}C-labelled PEG 4000 was promoted in PEG-grown cells and interaction of recombinant TonB-dependent receptor-like protein with PEG 200, 4000 and 20,000 strongly suggested that this protein is most probably a transporter of PEG into the periplasm [8]. A Gram-positive actinomycete, *Pseudonocardia* sp. strain K1, originally isolated as a tetrahydrofuran degrader, was also found to grow on PEG 4,000 and 8,000 [25]. Since this is the only Gram-positive bacterium known to grow on PEG, it would be interesting to see whether the actinomycete uses the same metabolic pathway and system for taking large PEGs into its cells. Pan and Gu [26] isolated Gram-negative and positive bacteria on PEG agar plates, but their viabilities in PEG liquid culture were null in 8 days and only 0.01-6% after 31 days. The colonization on agar plates is not always in accordance with the growth in PEG liquid culture. Since PEGs are finally going into the stream, the growth and degradation in PEG liquid medium or PEG-containing wastewater has to be most evaluated.

Symbiotic Biodegradation of PEG

Since the first report of PEG degradation by a symbiotic mixed culture [21, 27], many cases of symbiotic degradation of xenobiotic polymers have been reported with PEG-isophthalate copolymer, PVA, polyacrylate and polyaspartate [1].

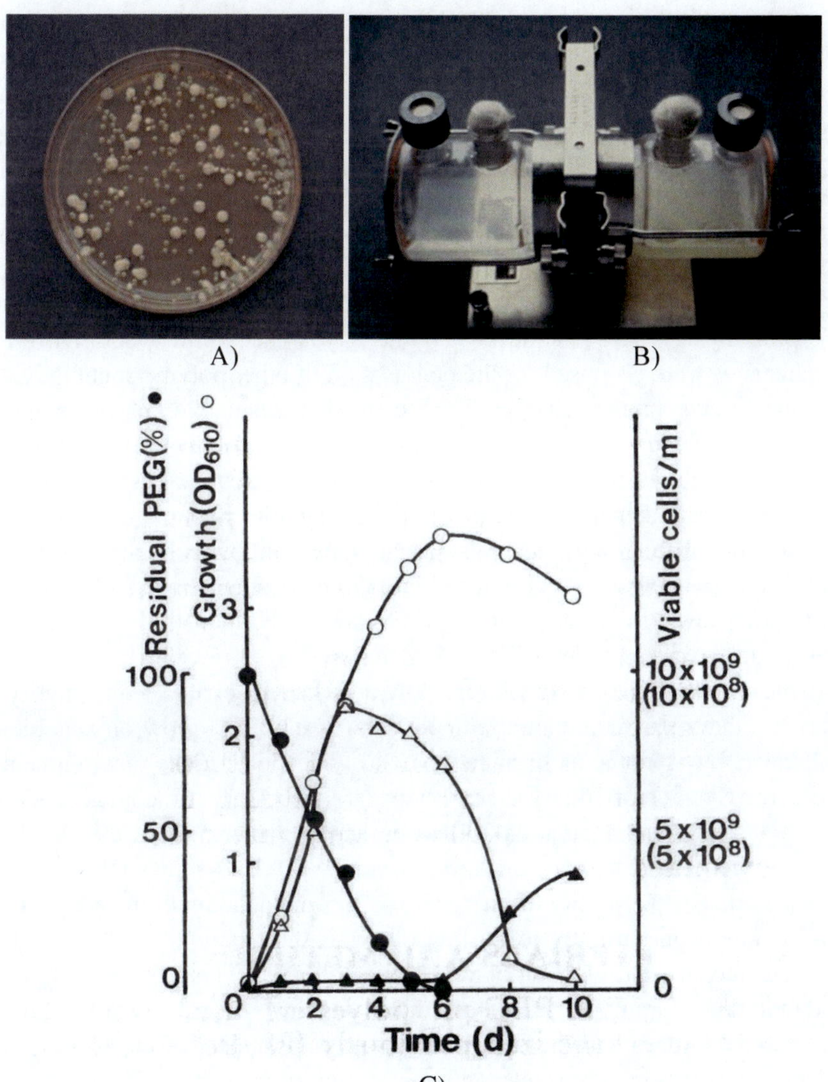

Figure 1. The symbiotic degradation of PEG by *sSphyngomonas terrae* and *ℓRhizobium* SP [1]. A, Mixed culture plate of two strains. Yellow colonies, *Sphingomonas terrae*; Wehite colonies, *Rhizobium* sp.: B, Dialysis culture of two strains in a Ecologen Model E-20 (New Brunswick Scientific). Right, *S. terrae*; Left, *Rhizobium* sp.: C, Growth of the mixed culture on PEG 6000. Symbols: open circle, growth; filled circle, residual PEG; open triangle, viable cells/ml of *S. terrae* indicated on a 10^9 scale; filled triangle, viable cells/ml of *Rhizobium* sp. indicated on a 10^8 scale.

Symbiotic degradation of PEG was reported with a PEG-assimilating mixed culture consisting of the PEG-degrader *Sphingomonas terrae*, which was reclassified from *Flavobacterium* sp. and a concomitant associate (*Rhizobium* sp., *Agrobacterium* sp., or *Methylobacterium* sp.) [20]. Although *S. terrae* is the main degrader and possesses all the enzymes necessary for the metabolism of PEG, it cannot grow on PEG as a single culture. Only mixed cultures of the strain with an associate (where *S. terrae* is dominant) show desirable growth on PEG (Figure 1) [27, 28].

The role of the concomitant associate in the symbiotic mechanism is based on detoxification of a toxic metabolite glyoxylate from PEG. Different mechanisms were proposed for the degradation of other polymers, but they are not the targets in this chapter. Pure cultures are suitable for biochemical elucidation of the metabolism or the mechanism of polymer degradation. In fact, pure cultures have been isolated in most of xenobiotics degradation. However, it is quite natural that microbial communities living in the ecosystem degrade the recalcitrant materials in the long run by collaborative work. Xenobiotic polymers were introduced into our ecosystem at most 80 years ago by the first invention of Nylon-66 by Carothers, WH. Industrial production of PEG with a variety of molecular sizes has started in 1950s [29]. Enzymes capable of metabolizing xenobiotic polymers had to evolve from prototype enzymes that recognize natural compounds to be able to grow on xenobiotic polymers by spontaneous mutation and selection under environmental pressure. Since more than a few steps are necessary to complete every metabolic pathway, adaptation had to be accomplished *via* all the steps. In addition, requirements for cofactors, optimum pH, and so on must also be considered. The possibility of one microbe accommodating all the adaptations and requirements is likely to be low over the short period of time that has elapsed since the introduction of xenobiotic polymers. This has resulted in the collaboration of multiple microbes (mixed culture). In other words, a huge variety of microbes have been preventing the environment of our planet from the fatal destruction, nevertheless a huge amount of chemical compounds have been industrially produced and a large portion of them have finally been released into the environment.

Molecular Mechanism in the Aerobic PEG Metabolism

Several reports have suggested different mechanisms in PEG degradation [5], but the most probable metabolic pathway is an exogenous metabolic one based on repeated oxidation steps, which is on line with the detection of carboxylated PEGs and shorter PEGs as the intermediate products [8, 9, 30]. Bernhard et al. [9] suggested that the biodegradation pathway of PEG 250 and 970 in seawater is similar to that for freshwater, but the degradation pathway for PEG 2,000-10,300 in seawater differs from the shorter PEGs, as no shortened PEGs were detected. However, the shift of the top peak to higher than the original PEGs meant that shorter components (and probably their products) were faster biodegraded. Detection of intermediate products is sometimes difficult, which is not against the oxidation mechanism commonly proposed for aerobic degradation of PEGs. PEG is oxidized by alcohol dehydrogenases linked with a dye or NAD [24]. PEG-dehydrogenase (PEG-DH) has been cloned from PEG-utilizing *S. terrae* and characterized as an FAD-including alcohol dehydrogenases [31, 32]. PEG-aldehyde dehydrogenase was cloned from a PEG-utilizing Sphingomonad and characterized as a NADP-containing nicotinoprotein PEG-aldehyde dehydrogenase [33], the first report of a nicotinoprotein aldehyde dehydrogenase. All the metabolic enzymes included in PEG degradation have been localized in the membrane and are thought to work in the periplasm, in accordance with the fact that PEG and its metabolites were detected in the periplasmic fraction [8], suggesting that PEG is taken up into the periplasm and metabolized there. We cloned the genes involved in PEG degradation, and found that the *peg* operon consisted of five genes and was expressed by PEG through induction of an *araC*-type regulator [34, 35], as shown in Figure 2. This was the first report on the regulation of degradative genes by a macromolecule. Two genes coding PEG-DH and PEG-aldehyde dehydrogenase are involved in the *peg* operon. A gene coding acyl-CoA synthetase in the *peg* operon was expressed and the recombinant enzyme was found to react with PEG-carboxylate to form PEG-carboxylate-CoA [36]. Judging from the nature of this kind of protein (located on the cytoplasmic membrane as a translocator), the enzyme might be responsible for the translocation of PEG-carboxylate from the periplasm into the cytoplasm or for the detoxification of strong acidity of the substrate. Since glutathione *S*-transferase (a gene coding this enzyme is located in the downstream of the *peg* operon) is localized in the cytoplasm and suggested to buffer the toxicity of PEG-carboxylate-CoA [37], the role of acyl-CoA synthetase might be to buffer the toxicity of PEG-carboxylate (perhaps a low molecular size) together with glutathione *S*-transferase. Other two genes code proteins related to transport, TonB-dependent receptor and permease respectively. TonB-dependent receptor forms the β-barrel structure and is localized in the outer membrane and its interaction with PEGs was suggested [8].

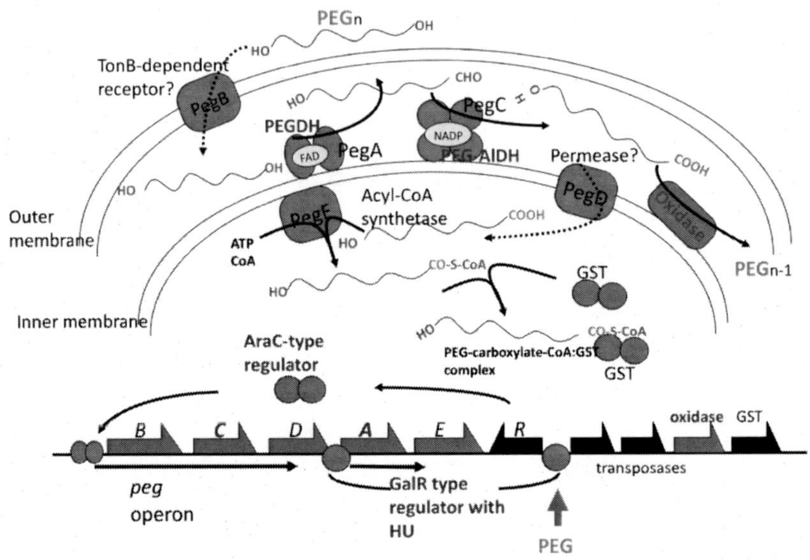

Figure 2. The operonic structure of genes involved in pegPEG degradation and its regulation by PEG [1].

Permease is homologous to glycoside cation symporters localized in the cytoplasmic membrane and might be a transporter for small PEG molecules or their metabolites produced in the periplasm. The ether bond-splitting enzyme involved in the PEG metabolism was perhaps a glycolic acid oxidase or glycolic acid dehydrogenase active on carboxylated PEG [38, 39]. A gene coding PEG-carboxylate dehydrogenase was detected in the downstream region of the *peg* operon, and was found to be involved in PEG metabolism [37]. In addition, superoxide dismutase cloned from PEG-utilizing *Pseudonocardia* sp. strain K1 showed ether-bond-cleaving activity for carboxylated PEG [40, 41].

These results and other reports [42] suggest that an ether bond is split by various types of enzymes, including monooxygenase, ether hydrolase, carbon-oxygen lyase, peroxidase, laccase, and glycolate oxidase in aerobic bacteria. This finding conflicts with the fact that metabolic enzymes in general evolved from an ancestor enzyme capable of recognizing a homologous chemical structure, and are therefore grouped into the same family.

Our group has already finished the analyses of the complete genomes of PEG-assimilating *S. macrogoltabida* strain 203 and *S. terrae* (they are the type strains of two species), which would appear in Genome Announcements (http://genomea.asm.org/) in the near future. The genome analysis of PEG-assimilating strains endoresed the presence and significance of the *peg* operon for the aerobic metabolism of PEG. Together with the complete genome analyses, recent rapid progress in transcriptome analysis would contribute to display the whole transcription relevant to the PEG metabolism. Hu et al. isolated various PEG degraders including not only Sphingomonads, but also other genera and found that all of them possessed a *pegA* gene [71]. The primers designed from this gene might be a good indicator to detect PEG degraders.

Anaerobic Biodegradation of PEG

Anaerobic biodegradation of PEG has been well investigated as compared to other polymers. Schink et al. reported that higher molecular mass PEG (up to PEG 40,000) was degraded by anaerobes isolates from limnic and marine muds [43, 44]. They proposed an anaerobic metabolic route for PEG, and suggested that acetaldehyde was produced *via* a hydroxyl shift reaction from PEG by a diol dehydratase-like enzyme (PEG acetaldehyde lyase) in the cytoplasmic space, but they failed to purify the enzyme [45]. Schink's group also demonstrated conversion of 2-phenoxyethanol to phenol and acetaldehyde in a way similar to a diol dehydratase reaction by a strictly anaerobic Gram-positive bacterium, *Acetobacterium* strain LuPhet1, but could not rule out an alternative pathway for the production of acetaldehyde [46]. Dwyer and Tiedje obtained a methanogenic consortium from sewage sludge that degraded ethylene glycol to PEG 20,000 [47]. Otal and Lebrato [48] confirmed the efficient removal of PEG mixtures (400-1000 or 1,500-10,000) with activated sludge from the anaerobic reactor of a wastewater treatment plant adapted to PEG. *Alcaligenes faecalis* var. *denitrificans* TEG-5, the first PEG-degrader able to degrade PEG under aerobic conditions [16], displayed PEG degradation under anaerobic nitrate-reducing conditions [49]. Microbes isolated from the wastewater treatment sludge decomposed PEGs up to 20,000 Da, either under aerobic or anaerobic conditions [10].

In general, the degradation rate under anaerobic conditions is lower than under aerobic conditions. Since most wastewater including activated sludge can be treated with aerobic systems, aerobic metabolic systems and aerobic microorganisms are very important for the degradation of xenobiotic polymers, but some fragments of the polymers enter anoxic environments after use. In addition, anaerobic treatment of wastewater increases its significance, as the anaerobic treatment produces methane (an energetic material), instead of CO_2 (a source of global warming).

Non-Metabolic Biodegradation of PEG

As described above, polymer-degrading microorganisms utilize target polymers in general as the sole carbon source, which possess the metabolic enzymes adapted to polymers or their metabolites. On the other hand, it is known that non-specific degradation of polymers is mediated by wood-rot fungi, just as found with lignin degradation by them. The first report on the degradation of synthetic polymeric materials by lignin-degrading fungi described the degradation of nylon-6 and -66 by white-rot fungi, including an isolate (IZU-154) and stock cultures [50]; nylon-degrading activity is based on oxidation of nylon by manganese peroxidase and is closely related to the lignolytic activity of fungi [51]. Larking et al. [52] found that the degradation of PVA was promoted by a combination of treatments with Fenton's reagent followed by biological degradation, probably by laccase produced in the culture supernatant of the white-rot fungus *Pycnoporus cinnabarinus*. Another lignin-degrading white-rot fungus, *Phanerochaete chrysoporium,* excreted lignin peroxidase, which promoted the degradation of PVA chains through the formation of carbonyl groups as well as double bonds [53]. A substantial decrease (approximately 80%) in average molecular mass was observed. Since the carbon chain of oxidized PVA is readily cleaved through non-enzymatic processes, strong oxidation of PVA either by an oxidizing enzyme or by a Fenton reaction leads to cleavage of the main carbon chain and yields oligomeric materials. Oligomers are generally biodegradable, as are oligomeric ethylene, styrene, and isoprene, although their polymers are non-biodegradable [54]. On the other hand, the brown-rot fungus *Gloeophyllum trabeum* secretes quinones that reduce Fe^{3+} and produce H_2O_2, resulting in an extracellular Fenton reaction degrading PEG [55]. Thus, brown- and white-rot fungi can play a significant role in the recycling of materials in the environment (especially in a forest or woods) and their degradation-related enzymes have potential applications in the treatment of polymer wastes and wastewater.

III. BIODEGRADATION OF POLYETHOXYLATED SURFACTANTS

As described above, the majority of PEGs are used as hydrophilic parts of nonionic surfactants, which are finally liberated into streams after use. Therefore, the biodegradation of nonionic PEG-containing surfactants is the environmentally serious issue in the aqueous environments such as streams, rivers, lakes, etc. As they are often contained in the wastewater from industries and homes, their biodegradability collected concerns of engineers managing industrial or municipal sewage treatment systems. In general, the ethoxy chain length included in nonionic surfactants is less than PEG 1000 and the most of surfactants has PEG 400 or so. As described above, this size of free PEG is rather readily degraded by a variety of bacteria. However, the ethoxy chain as a hydrophilic component combined with a hydrophobic component reduces the biodegradability. Aliphatic alcohol polyethoxylates and alkylphenol polyethoxylates are widely used as nonionic surfactants. Steber and Wierich reported the formation of carboxylated intermediates (carboxylated alcohol ethoxylates and carboxylated free PEGs) from stearyl alcohol heptaethoxylate treated with the activated sludge of a municipal sewage treatment plant [56]. Marcomini et al. [30] tested behavior of aliphatic alcohol polyethoxylates and identified free PEGs and carboxylated forms of the parent surfactant and free PEGs as intermediates from linear and monobranched aliphatic alcohol ethoxylates under standardized aerobic conditions. The central cleavage between linear or monobranched alcohol and ethoxy chain was suggested, which must have formed free PEGs, but mulchbranched alcohol polyethoxylates did not form free PEGs. In general, an ether bond is resistant for hydrolysis under neutral conditions, which is considered correct for ether bonds in PEGs, but the ether bond between a linear and monobranched aliphatic alcohol group and an ethoxy chain may be weak enough to be partially hydrolyzed. However, the hydrolysis is only partially, as carboxylaed ethoxylates of the original size were formed. Biodegradation of alcohol ethoxylate (AE; a mixture of 12 and 13 carbon length alkyl chains with an average of 5 ethylene oxide chain) indicated that free PEGs were readily degraded by AE-acclimated wastewater microbial consortium and the ether scission between alcohol and ethoxy chain was dominant in a wastewater community [57]. As ethylene glycol and diethylene glycol were not utilized by microbial community, although the community degraded higher molecular weight PEGs (up to octamer), they were thought to be accumulated. These results have suggested that the aerobic primary degradation of ethoxylate is the oxidation of a terminal alcohol group to a carboxyl group, as found with the aerobic metabolism of PEG in the preceding section. PEG degraders also can grow on some nonionic PEG-based surfactants [15, 58]. However, PEG-based surfactant degraders often cannot degrade free PEGs. Maki et al. suggested the carboxylation of alkylphenol ethoxylates by alkylphenol ethoxylate-degrading *Pseudomonas* sp. strain TR01, which can metabolize alcohol ethoxylates, but not PEG [59]. The same growth characteristics (growth on ethoxylates and no growth on PEGs) have been reported with *Pseudomonas* sp. RWA and RWX that were isolated as degraders for nonylphenol ethoxylate [NPEO] and *iso*-tridecyl ethoxylate, respectively [15]. The octylphenol ethoxylate (OPEO)-oxidizing dehydrogenase (OPEO-DH) purified from OPEO-degrading *Pseudomonas putida* S-5 also acted only on ethoxylates, but not on PEGs [60]. By the metabolism of NPEOs or OPEOs, nonylphenol diethoxylate (NP2EO) and its carboxylated form or octylphenol diethoxylate (OP2EO) were finally detected as the end product in the culture supernatants of NPEOs or OPEOs degraders [59, 61, 62]. Unlike linear or monobranched alcohol ethoxylates, the ether bond between alkylphenol and ethoxy chain does not appear to be hydrolyzed. It is known that shortened alkylphenol ethoxylate (diethoxylates or monoethoxylates; NP2EO /OP2EO or NP1EO/OP1EO) and alkylphenols are more recalcitrant to biodegradation. In the past, these intermediates were thought to be more toxic (estrogenic) than the original compounds, but recent research denied their effect unless the unusual high concentrations were used for experiments and no serious effect was found with mammals. Still the degradation of these intermediates is important, as the accumulation of any intermediates is undesirable in the environments. NP2EO and NP1EO, as well as their carboxylated compounds, were further transformed to nonylphenol (NP) under anaerobic conditions [63, 64]. On the other hand, Liu et al. isolated bacteria (*Ensifer* sp. AS08 and *Pseudomonas* sp. AS90) that can metabolize short ethoxy chain-nonylphenol (NPEO$_{av2.0}$: average 2.0 EO units) under aerobic conditions [65]. The terminal alcohol group was oxidized to a carboxylic acid. Then the carboxylated nonylphenol ethoxylate was depolymerized by one glycol unit, as found in the metabolism of free PEGs. This process was repeated until NP was finally produced. They cloned NPEO dehydrogenase (NPEO-DH: flavoprotein alcohol dehydrogenase) from *Ensifer* sp. strain As08, which can oxidize triethylene glycol ~ PEG 6000 and NPEOs [66] and elucidated the different activities of NPEO-DH, PEG-DH from *S. terrae* [31, 32] and OPEO-DH from *Pseudomonas putida* S-5 [60] on free PEGs and ethoxylates, based on the amino acid sequences and 3D modeling

[67]. Alignment of amino acid sequences of these enzymes has indicated that they comprises a PEGDH subgroup in the family of GMC oxidoreductases [68], but differences in the size, secondary structure and hydropathy of their active sites caused differences in the substrate specificities toward alkylphenol ethoxylates and free PEGs.

This has indicated that different growth on ethoxylates and free PEGs are caused by different preference of PEG-DH members toward substrates (free PEGs and ethoxylates; different polymerization degrees of EO; and their resultant hydropathy). It is quite natural that the preference is highly dependent on carbon sources with which the source bacteria were screened. Apart from these data, John and White [62] denied the carboxylation of ethoxylates and insisted that under aerobic and anaerobic conditions, nonylphenol ethoxylates are depolymerized similarly, liberating acetaldehyde. However, this has not been validated to date.

Linear alkyl ethoxylates (LAE) and PEGs were anaerobically degraded to methane and CO_2 by enrichment culture with LAE and pure cultures (*Peelobacter propionicus* strain KoB35 and *Acetobacterium* sp. strain KoB58) obtained from the enrichment culture [69]. Both strains produced fatty acids from LAE, completely degrading an ethoxy chain. No ether scission between alcohol and an ethoxy chain was suggested, but it may be probable, as suggested by aerobic degradation of LAE [30, 56].

In summary, it is obvious that alcohol ethoxylates and alkylphenol ethoxylates are metabolized aerobically *via* the carboxylation of the terminal alcohol group, as found with the metabolism of PEG. Alcohol ethoxylates appear to be partially hydrolyzed to alcohol and ethoxy chain under aerobic conditions by either a biotic or abiotic process and PEGs produced were carboxylated at the terminal alcohol groups. Alkylphenol ethoxylates are carboxylated without the ether scission between alkylphenol and ethoxy chain. Alkylphenol and its mono- or diethoxylates and their carboxylates were accumulated as the end products. All the alcohol dehydrogenases acting on PEGs and ethoxylates were elucidated at molecular levels and classified into the same PEG-DH subgroup in GMC flavoprotein oxidoreductase family, but they differ in the preference toward free PEGs and ethoxylates. On the other hand, alcohol ethoxylates were anaerobically degraded by enrichment culture and pure cultures, which depolymerized a PEG component due to the mechanism proposed for the anaerobic degradation of PEG and finally produced fatty acids from alcohols. Thus, the principal metabolisms of free PEGs and ethoxylates happen similarly at the terminal alcohol groups, either aerobic or anaerobic.

CONCLUSION

Degradation studies of PEG are one of pioneering works on biodegradability of xenobiotic polymers and biodegradation studies of xenobiotic compounds. Long history of research on biodegradation of PEGs up to 20,000-40,000 and polyethoxylated surfactants recorded that they are biodegradable under either aerobic or anaerobic conditions in freshwaters, seawaters or sewage treatment systems. The primary biodegradation starts at the terminal alcohol groups of either free PEGs or ethoxylates. Under aerobic conditions, the alcohol groups are oxidized to a carboxyl group and then carboxylated PEGs are depolymerized by one glycol unit. Enzymatic and genetic works have displayed unique enzyme groups; for example, PEG-DH group made a new branch in GMC flavoprotein oxidoreductases and PEG-aldehyde dehydrogenase was the first nicotinoprotein aldehyde dehydrogenase. The ether bond-splitting enzyme in the metabolism of PEG and other examples indicated that ether bond-splitting reactions are not catalyzed by ether bond-specific enzymes, but by various divergent enzymes such as monooxygenase, oxidase, dehydrogenase, hydrolase, and lyase [1]. Under anaerobic conditions, most probably acetaldehyde is produced *via* a hydroxyl shift reaction from the terminal alcohol groups by diol dehydratase-like enzyme, although no report on the purification of the enzyme is noted. The degradation of polymers was at first expected to be catalyzed solely by extracellular enzymes, given the assumption that macromolecules are never incorporated into cells, but this was disproved by the periplasmic degradation of PEG and PVA. These macromolecules are surely incorporated into the periplasm through the outer membranes of Sphingomonads and are metabolized by periplasmic enzymes, although the mechanism of macro-molecule uptake has not been sufficiently characterized yet. Putative diol dehydratase-like enzyme relevant to anaerobic PEG degradation was concluded to be localized in the cytoplasmic fraction, which also suggested the uptake of PEG into the cells. A PEG-DH gene from Sphingomonads has been distributed and conserved among different genera in the 40 years since PEG-degrading Sphingomonads were isolated in 1975 and a gene for PEG-DH was cloned in 2001, suggesting a significant role of a large plasmid harboring a PEG-degradative gene cluster in circulation of degradation ability among microorganisms. Since a *pva* operon is also on a large plasmid of *Sphingopyxis* sp. strain 113P3, the operon structure or involved enzyme genes have probably been distributed among microorganisms. These findings strongly suggested that the short history of xenobiotic polymers to date has been sufficient for degrading enzymes to evolve from the prototype enzymes and for the microorganisms (pure culture or consortium) to adapt to an environment contaminated by xenobiotics. In addition, megaplasmids must have sped up the distribution rate of degrading ability. The existence of Sphingomonads that degrade various xenobiotic polymers, such as PEG and PVA, *via* intracellular enzymes suggest that they have a means of taking macromolecules into the periplasm as well as metabolic enzymes adapted to these macromolecules. Several Psudomonads were isolated as high molecular mass-PEG degraders. As some Pseudomonads were taxonomically reidentified as a member of *Sphingomonas* [70], these Pseudomonads might be Sphingomonads, when based on the recent taxonomy. Unfortunately no further work was documented and no strain was deposited to stock culture collections except Sphingomonads.

It is notable that the principal degradation of ethoxy surfactants is performed as in the parent form, based on the mechanism similar to that of free PEGs, but aliphatic (either linear or monobranched) alcohol ethoxylates are partially cleaved to free PEG and a hydrophobic component, in which PEG and a hydrophobic component are metabolized, independently. Aliphatic alcohol and alkyphenols occupy the major hydrophobic components in ethoxy surfactants. The chain length of aliphatic alcohols is not long and the alcohols are metabolites of short alkanes, which are further oxidized to fatty acids and readily metabolized by β-oxidation. In the aquatic environments, alkylphenols can be partially removed by volatilization from the waters and biodegradation of alkylphenols has been reported [11].

The many examples of symbiotic polymer degradation appear to suggest that symbiotic degradation occurs in an ecosystem when new artificial compounds are introduced. In fact, biodegradation in sewage treatment plants is symbiotic. At the same time, non-metabolic polymer degradation suggests that the ecosystem has the versatile strategies in terms of disposal of new artificial compounds.

Enzymes in general catalyze the reaction of target materials by their substrate specificities and the catalytic properties of them. These enzymes are metabolizing enzymes, which are found in most cases of polymer degradation. However, strong oxidative reaction can attack materials non-specifically and cleave them into small fragments. This is not relevant to the metabolism of materials. Wood-rot fungi are known to excrete powerful oxidizing enzymes, which attack non-specifically many materials. As most of PEG and ethoxy surfactants eventually go into the aquatic environments, the role of wood-rot fungi does not appear to be significant in the ecosystem. However, the potential of wood-rot fungi might be useful for specific cases that `PEGs and ethoxy surfactans are not entering the streams.

Since water-soluble polymers are neither incinerated nor recycled after use, microbial degradation is the only means of cleaning contaminated environments. From the distribution of degrading microorganisms and the rate of degradation, it is possible to predict the fate of target compounds in the environment. Even if target polymers are nontoxic, they have strong surface activity and produce large amounts of foam, preventing the recovery of oxygen in the water. This poses a serious threat to water-borne organisms as well as to humans. With regards to PEGs and alkylphenol ethoxylates, ethylene oxide oligomers, mono or diethoxy alkylphenols and their carboxylates are more toxic and recalcitrant than the parent structures, although these metabolites appear to be biodegraded in the long run and cause no serious and fatal effect on the ecosystem. Therefore, the application of degrading microorganisms to wastewater treatment is important and expected to expand in the future.

ACKNOWLEDGMENTS

The author is grateful to many scientists and graduates, assisting technicians and secretaries for their collaboration, help and assistance during the research on the topics described here. She also acknowledges the financial supports from the ministry of Education, Culture, Sports, Science and Technology, Japan and many private foundations.

REFERENCES

[1] Kawai, F, The biochemistry and molecular biology of xenobiotic polymer degradation by microorganisms. *Biosci. Biotechnol. Biochem.*, 2010. 89, 1743-1759.

[2] Roy AB; Curtis CG; Powell GM. The metabolic sulphation of polyethylene glycols by isolated perfused rat and guinea-pig livers. *Xenobiotica,* 1987. 17, 725-732.

[3] Herald, DA; Keil, K; Bruns, DE. Oxidation of polyethylene glycols by alcohol dehydrogenase. *Biochem. Pharmacol.*, 1989. 38, 73-76.

[4] Gordienko, AD; Kudokotseva, EV. Study of the functional composition of mitochondria in cellular suspensions subject to a cryoprotectant solution, *Kriobiol. Kriomed.*, 1980. 7, 32-34.

[5] Kawai, F. Microbial degradation of polyethers. *Appl. Microbiol. Biotechnol.*, 2002. 58, 30-38.

[6] Cox, DP. The biodegradation of polyethylene glycols. In: Perlman D. editor. *Advances in Applied Microbiology*, Vol. 23. New York: Academic Press; 1978: 173-194.

[7] Harris, JM. Poly(ethylene glycol) Chemistry, Edition. New York: Plenum Press: 1992.

[8] Kawai, F; Kitajima, S; Oda, K; Higasa, T; Charoenpanich, J; Hu, X; Mamoto, R. Polyvinyl alcohol and polyethylene glycol form polymer bodies in the periplasm of Sphingomonads that are able to assimilate them. *Arch. Microbiol.*, 2013. 195, 131-140.

[9] Bernhard, M; Eubeler, JP; Zok, S; Knepper, TP. Aerobic biodegradation of polyethylene glycols of different molecular weights in wastewater and seawater. *Water Res.*, 2008. 42, 4791-4801.

[10] Huang, Y-L; Li, Q-B; Deng, X; Lu, Y-H; Liao, X-K; Hong, M-Y; Wang, Y. Aerobic and anaerobic biodegradation of polyethylene glycols using sludge microbes. *Process Biochem.*, 2005. 40, 207-211.

[11] Jonkers, N; de Voogt, P. Non-ionic surfactants in marine and estuarine environments. In: Knepper, TP; Barcelø, D; de Voogt, P. editors. Comprehensive analytical Chemistry, Vol. XL. Amsterdam; Elsevier *Science BV;* 2003; 719-747.

[12] Zgola-Grzeskowiak, A; Grzeskowiak, T; Zembrzuska, J; Lukaszewski, Z. Comparison of biodegradation of poly(ethylene glycol)s and poly(propylene glycol)s. *Chemosphere*, 2006. 64, 803-809.

[13] Abel, M; Conrad, T; Giger, W. Persistent organic chemicals in sewage effluents. 3. Determination of nonylphenoxycarboxylic acids by high-rcsolution gas chromatography/mss spectrometry and high-performance liquid chromatography. *Environ. Sci. Technol.*, 1987. 21, 697-703.

[14] Reinhard, M; Goodman, N; Mortelmans, KE. Occurrence of brominated alkylphenol polyethoxy carboxylates in mutagenic wastewater concentrates. *Environ. Sci. Technol.*, 1982. 16, 351-362.

[15] Corti, A; Antone, SD; Solaro, R; Chiellini, E. Degradation of poly(ethylene glycol)-based nonionic surfactants by different bacterial isolates from river water. J. Environ. Polym. Degr., 1998. 6, 121-131.

[16] Payne, WJ. Pure culture studies of the degradation of detergent compounds, *Biotechnol. Bioeng.*, 1965. 5, 355-365.

[17] Haines, J; Alexander, M. Microbial degradation of polyethylene glycols. *Appl. Microbiol.*, 1975. 29, 621-625.

[18] Hosoya, H; Miyazaki, N; Sugisaki, Y; Takanashi, E; Tsurufuji, M; Yamasaki, M; Tamura, G. Bacterial degradation of synthetic polymers and oligomers with the special reference to the case of polyethylene glycol. *Agric. Biol. Chem.*, 1978. 42, 1545-1552.

[19] Obradors, N; Aguilar, J. Efficient biodegradation of highmolecular-weight polyethylene glycols by pure cultures of P*seudomonas stutzeri*. *Appl. Environ. Microbiol.*, 1991. 57, 2383-2388.

[20] Takeuchi, M; Kawai, F; Shimada, Y; Yokota, A. Taxonomic study of polyethylene glycol-utilizing bacteria;emended description of the genus *Sphingomonas* and new description of *Sphingomonas macrogoltabidus* sp. nov., *Sphingomonas sanguis* sp. nov. and *Sphingomonas terrae* sp. nov. *Syst. Appl. Microbiol.*, 1993. 16, 227-238.

[21] Ogata, K; Kawai, F; Fukaya, M; Tani, Y. Isolation of polyethylene glycols-assimilable bacteria. *J. Ferment. Technol.*, 1975. 53, 757-761.

[22] Kawai, F. Sphingomonads involved in the biodegradation of xenobiotic polymers. *J. Ind. Microbiol. Biotechnol.*, 1999. 23, 400-407.

[23] Yasuda, M; Chereanov, A; Duine, JA. Polyethylene glycol dehydrogenase activity of Rhodopseudomonas acidophila derives from a type I quinohaemoprotein alcohol dehydrogenase. *FEMS Microbiol. Lett.*, 1996. 138, 23-28.

[24] Kawai, F; Yamanaka, H. Inducible or constitutive polyethylene glycol dehydrogenase involved in the aerobic metabolism of polyethylene glycol. *J. Ferment. Bioeng.*, 1989. 67, 300-302.

[25] Kohlweyer, U; Thiemer, B; Schraeder, T; Andreesen, JR. Tetrahydrofuran degradation by a newly isolated culture of Pseudonocardia sp.strain K1. *FEMS Microbiol. Lett.,* 2000. 186, 301-306.

[26] Pan, L; Gu, J-D. Characterization of aerobic bacteria involven in degrading polyethylene glycol (PEG)-3400 obtained by plating and enrichment culture techniques. *J. Polym. Environ.*, 2007. 15: 57-65.

[27] Kawai, F; Yamanaka, H. Biodegradation of polyethylene glycol by symbiotic mixed culture (obligate mutualism). *Arch. Microbiol.*, 1986. 146, 125-129.

[28] Kawai, F. Bacterial degradation of a new polyester, polyethylene glycol-phthalate polyester. *J. Environ. Polym. Degrad.*, 1996. 4, 21-28.

[29] Kawai, F. Biodegradation of polyethers (polyethylene glycol, polyprpylene glycol, polytetramethylene glycol and others). In

Matsumura, S; Steinbüchel, A. editors. Biopolymers, Vol. 9. Weinheim: *WILEY-VCH*; 2003; 267-298.

[30] Marcomini, A; Zanette, M; Poiana, G; Suter, MJ-F. Behaviour of aliphatic alcohol polyethoxylates and their metabolites under standardized aerobic biodegradation conditions. *Environ. Toxicol. Chem.*, 2000. 19, 549-554.

[31] Sugimoto, M; Tanabe, M; Hataya, M; Enokibara, S; Duine, JA; Kawai, F. The first ste in polyethylene glycol degradation by Sphingomonads proceeds via a flavoprotein (FAD-containing) alcohol dehydrogenase. *J. Bacteriol.*, 2001. 183 (22), 6694-6698.

[32] Ohta, T; Kawabara, T; Nishikawa, K; Tani, A: Kimbara, K; Kawai, F. Analysis of amino acid residues involved in catalysis of polyethylene glycol dehydrogenase from Sphingopyxis terrae, using three-dimensional molecular modeling-based kinetic characterization of mutants. *Appl. Environ. Microbiol.*, 2001. 72. 4388-4396.

[33] Ohta, T; Tani, A; Kimbara, K: Kawai, F. A novel nicotinoprotein aldehyde dehydrogenase involved in polyethylene glycol degradation. *Appl. Microbiol. Biotechnol.*, 2005. 68, 639-646.

[34] Charoenpanich, J; Tani, A; Moriwaki, N; Kimbara, K; Kawai, F. Dual regulation of a polyethylene glycol degradative operon by AraC-type and GalR-type regulation in *Sphingopyxis macrogoltabida* strain 103, *Microbiology,* 2006. 152. 3025-3034.

[35] Tani, A; Charoenpanich, J; Mori, T; Takeuchi, M; Kimbara, K; Kawai, F. Structure and conservation of a polyethylene glycol-degradative operon in sphingomonads. *Microbiology,* 2007. 153, 338-346.

[36] Tani, A; Somyoonsap, P; Minami, T; Kimbara, K; Kawai, F. Polyethylene glycol (PEG)-carboxylate-CoA synthetase is involved in PEG metabolism in *Sphingopyxis macrogoltabida* strain 103. *Arch. Microbiol.*, 2008. 189, 407-410.

[37] Somyoonsap, P; Tani, A; Charoenpanich, J; Minami, T; Kimbara, K; Kawai, F. Involvement of PEG-carboxylate dehydrogenase and glutathione *S*-transferase in PEG metabolism by *Sphingopyxys macrogoltabida* strain 103. *Appl. Microbiol. Biotechnol.*, 2008. 81, 473-484.

[38] Yamanaka, H; Kawai, F. Purification and characterization of a glycolic acid (GA) Oxidase active toward diglycolic acid (DGA) produced by DGA-utilizing Rhodococcus sp.No.432. *J. Ferment. Bioeng.,* 1991. 71, 83-88.

[39] Enokibara, S; Kawai, F. Purification and characterization of an ether bond-cleaving enzyme involved in the metabolism of polyethylene glycol. *J. Ferment. Bioeng.*, 1997. 83, 549-554.

[40] Yamashita, M; Tani, A; Kawai, F. Cloning and expression of an ether-bond-cleaving enzyme involved in the metabolism of polyethylene glycol. *J. Biosci. Bioeng.*, 2004. 98, 313-315.

[41] Yamashita M; Tani A; Kawai F. A new ether bond-splitting enzyme found in Gram-positive polyethylene glycol 5000-utilizing bacterium, *Pseudonocardia* sp. strain K1, *Appl. Microbiol. Biotechnol.*, 2005. 66, 174-179.

[42] White GF; Russell NJ; Tidswell E, Bacterial scission of ether bonds. *Microbiol. Rev.*, 1996. 60, 216-232.

[43] Schink B; Stieb M. Fermentative degradation of polyethylene glycol by a strictly anaerobic,Gram-negative,non-sporeforming bacterium, *Pelobacter venetianus* sp.nov. *Appl. Environ. Microbiool.*, 1983. 45, 1905-1913.

[44] Strass A; Schink B. Fermentation of polyethylene glycol via acetaldehyde in Pelobacter venetianus. Appl Microbiol Biotechnol., 1986. 25, 37-42.

[45] Frings J; Schramm E; Schink B. Enzymes involved in anaerobic polyethylene glycol degradation by *Pelobacter venetianus* and *Bacteroide*s. *Appl. Environ. Microbiol.*, 1992. 58, 2164-2167.

[46] Speranza G; Mueller B; Orlandi M; Morelli CF; Manitto P; Schink B. Mechanism of anaerobic ether cleavage: conversion of 2-phenoxyethanol to phenol and acetaldehyde be *Acetobacterium* sp. *J. Biol. Chem.*, 2002. 277, 11684-11690.

[47] Dwyer DF; Tiedje JM. Metabolism of polyethylene glycol by two anaerobic bacteria, *Desulfovibrio desulfuricans* and a *Bacteroides* sp. *Appl. Environ. Microbiol.*, 1986. 52, 852-856.

[48] Otal, E; Lebrato, J. Anaerobic degradation of polyethylene glycol mixtures. *J. Chem. Technol. Biotechnol.*, 2003. 78, 1075-1081.

[49] Grant, MA; Payne WJ. Anaerobic growth of *Alcaligenes faecalis* var. *denitrificans* at the expense of ether glycols and nonionic detergents. *Biotechnol. Bioeng.*, 1983. 25, 627-630.

[50] Deguchi, T.; Kakezawa, M.; Nishida, T. Nylon biodegradation by lignin-degrading fungi. *Appl. Environ. Microbiol.*, 1997. 63, 329-331.

[51] Deguchi, T; Kitaoka, Y; Kakezawa, M; Nishida, T. Purification and characterization of a nylon-degrading enzyme. *Appl. Environ. Microbiol.*, 1998. 64, 1366-1371.

[52] Larking, DM; Crawford, RJ; Christie, GBY; Linergan, GT. Enhanced degradation of polyvinyl alcohol by *Pycnoporus cinnabarinus* after pretreatment with fenton's reagent. *Appl. Environ. Microbiol.*, 1999. 65, 1798-1800.

[53] Mejia, AI; Lucy Lopez, BL; Mulet, A. Biodegradation of poly(vinyl alcohol) with enzymatic extracts of *Phaenerochaete chrososporium, Macromol. Symp.*, 1999. 148, 131-147.

[54] Kawai, F. Breakdown of plastic and polymers by microorganisms. In Fichter, A. editor. Advances in biochemical engineering/biotechnology, vol. 52, Heidelberg: Springer; 1995; 151-194.

[55] Jensen Jensen, KA Jr.; Houman, CJ; Ryan, ZC; Hammel, KE. Pathways for extracellular Fenton chemistry in the brown rot basidiomycete *Gloeophyllum trabeum, Appl. Environ. Microbiol.*, 2001. 67, 2705-2711.

[56] Steber, J; Wierich, P. Metabolites and biodegradation pathways of fatty alcohol ethoxylates in microbial biocenoses of sewage treatment plants. *Appl. Environ. Microbiol.*, 1985. 49, 530-537.

[57] Sharvelle, SE; Garland J; Banks MK. Biodegradation of polyalcohol ethoxylate by a wastewater microbial consorcium. *Biodegradation*, 2008. 19, 215-221.

[58] Kawai F, Fukaya M, Tani Y, Ogata K. Identification of polyethylene glycols (PEGs)-assimilable bacteria and culture characteristcs of PEG 6000 Degradation by a Mixed Culture. *J. Ferment. Technol.*, 1977. 55 (5), 429-435.

[59] Maki, H; Masuda, N; Fujiwara, Y; Ike, M; Fujita, M. Degradation of alkylphenol ethoxylates by *Pseudomonas* sp. strain TR01. *Appl. Environ. Microbiol.*, 1994. 50 (7), 2265-2271.

[60] Tasaki, Y; Yoshikawa, H; Tamura, H, Isolation and characterization of an alcohol dehydrogenase gene from the octylphenol polyethoxylate degrader *Pseudocmonas putida* S-5. *Biosci. Biotechnol. Biochem.*, 2006. 70 (8), 1855-1863.

[61] Nishio, E; Ichiki, Y; Tamura, H; Morita, S; Watanabe, K; Yoshikawa, H. Isolation of bacterial strains that produce the endocrine disruptor, octylphenol diethoxylate, in paddy fields. *Biosci. Biotechnol. Biochem.*, 2002. 66(9), 1792-1798.

[62] John DM; White, GF. Mechanism for biotransformation of nonylphenol polyethoxylates to xenoestrogens in *Pseudomonas putida. Appl. Environ. Microbiol.*, 1985. 49 (3), 530-537.

[63] Giger, W; Brunner, PH; Schaffner, C. 4-Nonylphenol in sewage sludge: accumulation of toxic metabolites from non-ionic surfactants. *Science*,1984. 22, 623-625.

[64] Ejlertsson, J; Nilsson, ML; Kylin, H; Bergman, A; Karlson, I; Oquist, M; Svensson, BH. Anaerobic degradation of nonylphenol mono- and di-ethoxylates in digestor sludge, landfill municipal solid waste and landfill sludge. *Environ. Sci. Technol.*, 1999. 33, 301-306.

[65] Liu, X; Tani, A; Kimbara, K; Kawai, F. Metabolic pathway of xenoestragenic short ethoxy chain-nonylphenol to nonylphenol by aerobic bacteria, *Ensifer* sp. strain AS08 and *Pseudomonas* sp. strain AS90. *Appl. Microbiol. Biotehchnol.*, 2006. 72, 552-559.

[66] Liu, X; Tani, A; Kimbara, K; Kawai, F. Xenoestrogenic short ethoxy chain nonylphenol is oxidized by a flavoprotein alcohol dehydrogenase from *Ensifer* sp. strain AS08. *Appl. Microbiol. Biotechnol.*, 2007. 73, 1414-1422.

[67] Liu, X; Ohta, T; Kawabata, T; Kawai, F. Catalytic mechanism of short ethoxy chain nonylphenol dehydrogenase belonging to a polyethylene glycol dehydrogenase group in the GMC oxidoreductase family. *Int. J. Mol. Sci.*, 2013. 14, 1218-1231; doi: 10.3390/ijms14011218.

[68] Zämocky, M; Hallberg, M; Ludwig, R; Divne, C; Haltrich, D. Ancestral gene fusion in cellobiose dehydrogenases reflects a specific evolution of GMC oxidoreductases in fungi. *Gene,* 2004. 338, 1-14.

[69] Wagener, S; Schink, B. Fermentative degradation of noionic surfactants and polyethylene glycol by enrichment cultures and by pure cultures of homoacetogenic and propionate-forming bacteria. *Appl. Environ. Microbiol.,* 1998. 54, 561-565.

[70] Yabuuchi, E; Yano, I; Oyyaizu, H; Hashimoto, Y; Ezaki, T; Yamamoto, H. Proposals of iSphingomonas paucimobilis *gen. nov. and comb. Nov.,* Sphingomonas parapaucimobilis *sp. nov.,* Sphingomonas yanoikuyae *sp. nov.,* Sphingomonas adhaesiva *sp. nov.,* Sphingomonas capsulata *comb. Nov.,* and two genospecies of the genus Sphingomonas. Microbiol. Immunol., 1990. 34, 99-119; *Int. J. Syst. Bacteriol.,* 1990. 40, 320-321.

[71] Hu, X; Fukutani, A; Liu, X; Kimbara, K; Kawai, F. Isolation of bacteria able to grow on both polyethylene glycol (PEG) and polypropylene gycol (PPG) and their PEG/PPG dehydrogenases. *Appl. Microbiol. Biotechnol.,* 2007. 73, 1407-1413.

In: Polyethers and Polyethylene Glycol   ISBN: 978-1-63483-392-9
Editor: Joyce Williams   © 2015 Nova Science Publishers, Inc.

Chapter 3

POLYETHYLENE GLYCOL BASED PHASE CHANGE POLYMERS FOR THERMAL ENERGY STORAGE APPLICATION

A. B. Samui[*]

Dept. of Applied Chemistry,
Defence Institute of Advanced Technology (Deemed University),
Pune, India

ABSTRACT

Phase change material can make a great contribution to the energy saving by way of energy storage in the form of latent heat. It can act both ways as when temperature is to be reduced it melts from crystalline phase while temperature of the substrate is prevented from rising. Similarly, during heating, a pre-molten phase of a crystalline material releases the heat while getting crystallized. In the process further cooling of the substrate is prevented.

A series of strategies are incorporated in the chapter and their heat storage details are discussed. As the phase change material is polyethylene glycol (PEG) it has either to be used by constraining in microcapsule or to be prevented from flowing during melting by adopting other methodologies for convenient use.

The first strategy discussed is the form-stable composition which does not allow PEG to flow during melting due to various secondary forces restricting their separation. This strategy is practiced by physically

---

[*] Corresponding author: e-mail: absamui@gmail.com; asit_samui@yahoo.com.

mixing PEG with other polymers or porous carbons, silica, nanoclay and others. The process advantage is further enhanced by mixing some materials which enhance the thermal conductivity ensuring fast action.

Chemically reacted PEG copolymers and others are always non-flowable although the enthalpy may be sacrificed.

PCM fibers and textiles are also discussed. In both form stable, microcapsulated PCM adherence to textile as well as direct chemical linking of PCM with textile have been discussed.

Lastly, the incorporation in concrete, building structure, PU foam etc. are discussed and their advantages and disadvantages are discussed for better understanding.

The characterization and performance study are also incorporated in the chapter.

The non-renewal energy sources are diminishing rapidly as the need of human is increasing constantly. Therefore, the most urgent need now is to maximize the renewable energy resources. Industrial developments and population boom in the past few centuries have added enormous pressure on the increasing energy demand. The Outlook predicted by considering several factors that global energy consumption will rise by 41 per cent from 2012 to 2035 [1].

There is always a requirement of thermal management of commonly used electronic devices like cell phones, laptop, computers, antennae and similar appliances and so on. Many research studies are directed towards improvement of thermal management, as the size of electronic equipment can become smaller with efficient power dissipation ability. This will minimize the use of space consuming standard cooling system dependent on electrical power.

Considering above and the use related environmental pollution affecting mass and climate, the research work on solar energy, hydropower, wind energy, geothermal energy, ocean energy, bioenergy, chemical energy etc. took the center stage. In fact several technologies have reached maturity and are being exploited with success.

Under this background one more concept creeps in mind about the possibility of reduction in power consumption to keep the demand lower. Among only few concepts in this direction, the only effective method is thermal energy storage, which can be used cleverly to put lesser load on energy demand. Thermal energy can be stored by adopting various procedures such as [2]:

- Sensible Thermal Energy Storage
- Underground Thermal Energy Storage (UTES)
- Thermal Energy Storage via Chemical Reactions
- Phase Change Materials (PCM)

UTES and TES via chemical reaction involve a number of complicated processes. Sensible heat storage is relatively inexpensive, but its drawbacks are its low energy density and its variable discharging temperature [3]. All these issues can be taken care by phase change materials. The PCM-based TES, which enables higher storage capacity, has other attribute in the temperatures range selection as desired. The change of phase could be either a solid/liquid or a solid/solid process. Melting processes involve energy densities of about 100 kWh/m^3 (e.g., ice), compared to a typical 25 kWh/m^3 for sensible heat storage options. In fact the idea of using PCM in a storage tank has been developed in the 80s with paraffin. Later paraffin was found to have a major drawback in the form of its flammability.

Although it is mostly applied for cooling and also in few cases of heating, there are possibilities of use in various technologies. Potential application areas of PCM found from investigations by Fatih Dermirbas [4] are described as:

- Thermal protection of flight data and cockpit voice recorders
- Hot and cold medical therapy
- Transportation and storage of perishable foods, medicine and pharmaceuticals products
- Solar power plants to store thermal energy during day time and reuse it during the latter part of the day
- Electronic chips to prevent operation at extreme temperatures
- Photovoltaic cells and solar collectors to avoid hot spots
- Miscellaneous use like solar-activated heat pumps, waste heat recovery etc.

The simplest PCMs are materials which utilize latent heat absorbed or released over a narrow temperature range of the material undergoing phase transformation. These materials absorb heat from the surroundings and melt during the heating process and release the heat back to the surroundings by crystallization during the cooling process as shown in scheme 1. Different PCMs differ in the phase change temperature and the heat storage capacities.

The requirements of these materials are as follows [5]:

- Large phase change enthalpy (>100kJ/kg)
- Suitable phase change temperature
- Reproducible phase change cycle
- Good thermal conductivity (~0.5W/mK)
- Low cost
- Chemical stability of PCM
- Non-toxic

The storage capacity of latent heat thermal energy storage can be given by the following equation [6]:

$$Q = \int_{Ti}^{Tm} mC_p dT + ma_m \Delta h_m + \int_{Tm}^{Tf} mC_p dT$$

$$Q = m\left[C_{sp}(T_m - T_i) + a_m \Delta h_m + C_{lp}(T_f - T_m)\right]$$

Where Q is the storage capacity, C_p specific heat, T_i, T_m, and T_f are initial, melting and freezing temperature respectively and h is the enthalpy

COMMONLY USED PHASE CHANGE MATERIALS

Both inorganic and organic class of materials suit the requirement of PCM.

Hydrated Salts

Hydrated salts consist of salt bound to water molecules in a fixed proportion. They have high storage density per unit volume and relatively high thermal conductivity, but there are problems like supercooling and phase segregation on repeated cycling. The hydrated salt, $CaCl_2.6H_2O$ is one of the widely used inorganic PCM.

Eutectic Mixtures

Eutectics are mixtures of two or more components with a minimum freezing point. It offers the advantage of no phase segregation and shows congruent melting. For example: a eutectic mixture of 66.6% $CaCl_2.6H_2O$ and 33.3% $MgCl_2.6H_2O$ melts at 25°C having phase change enthalpy of 127kJ/kg.

Paraffin

Paraffin waxes such as n-eicosane, n-heptadecane are aliphatic linear alkanes with the general formula C_nH_{2n+2}. It has large thermal storage capacity and is easily available at low cost.

Fatty Acids

Fatty acids have the general formula of $CH_3(CH_2)_nCOOH$. They have good thermal storage capacity which increases with increasing length of alkyl chain. There is no phase separation and they are stable to cycling. For instance, stearic acid has a melting point of 66.8°C with 258.98 J/g latent heat of fusion.

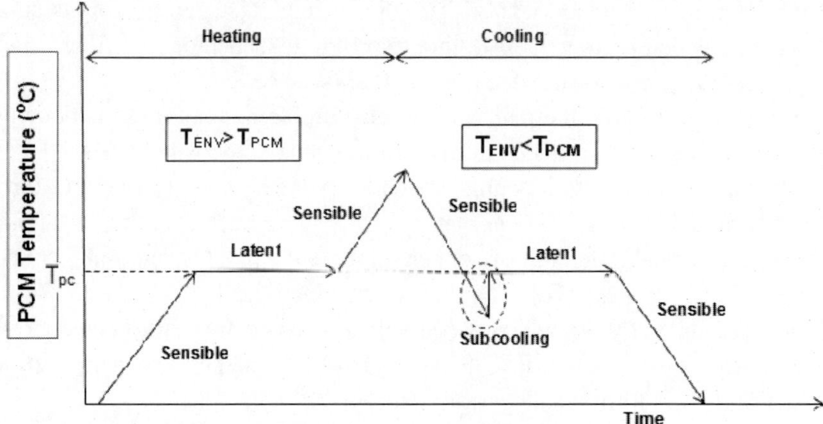

Scheme 1. Schematic representation of functioning of PCM.

Polyethylene Glycol

Polyethylene glycol has the basic unit of $-CH_2-CH_2-O-$. It has the superior characteristics of high latent heat capacity, wide temperature range, ease of modification, congruent melting and crystallizing behavior, non-corrosive and non-toxic in nature. The melting temperature and enthalpy are dependent on the molecular weight of PEG.

The organic PCMs have been considered as having a good potential for application in thermal energy storage and intense research activity on organic PCMs followed [7, 8].

However, the basic deficiencies inherent in organic PCMs, such as the leakage in the molten state and lower thermal conductivity (around 0.2 W/m K) restricted their wide spread application. Therefore, various methods have been adopted to overcome these problems. The most prospective and practically useful method is the design of form-stable composite PCMs [9].

The PCMs like PEG, fatty acids and paraffin waxes can be dispersed into a higher melting point polymer which acts as the supporting material. With increasing temperature, the supporting material retains its shape even after the PCM has melted. The transition is a solid-liquid phase change. But the PCM is embedded in such a manner in the polymer matrix that there is no leakage. This is known as form-stable or shape stabilized PCM. This process is gaining significant attention [10].

The porous inorganic materials or silicate minerals such as porous silica [11, 12], expanded perlite [13, 14], montmorillonite or bentonite [15] etc. have been used as supports to stabilize organic PCMs.

To address the low thermal conductivity problem, some high conductive materials are usually introduced into the form-stable composite PCMs.

The thermal conductivity enhancement of polyethylene glycol (PEG)/SiO_2 hybrid form-stable phase change material (PCM) was achieved by Cu doping via in situ chemical reduction of copper sulphate ($CuSO_4$) through ultrasound-assisted sol–gel process [16]. The thermal conductivity is 0.414 W/(m K) for 2.1 wt% Cu in PEG/SiO_2, which is equivalent to about 38% enhancement over the undoped one . The Cu/PEG/SiO_2 hybrid material has excellent thermal stability and good form-stable performance.

The short and long multi-walled carbon nanotubes, carbon nanofibers, and graphene nanoplatelets (GNPs) were used for the purpose by Fan et. al. [17]. Nanocomposite PCM samples were prepared with each type of nanofillers having mass concentrations of 1–5 wt.%. The melting enthalpies were slightly reduced while melting temperature remains mostly unaffected. The thermal conductivity increases with loading although size and shape decide the extent of increase. The GNPs loading of 5 wt% shows an increase of about 164%.

The expanded graphite was added to the composite to enhance the thermal conductivity of the composite and the increase was found to be high. In a very recent report utilization of similar material was described for making form stable PCM as well as for enhancing thermal conductivity [18]

Graphene oxide (GO) sheets were introduced to stabilize the shape of the most widely studied phase change material, PEG during the solid–liquid phase change process. The shape-stable PCM presents a high heat storage capacity of 142.8J/g and excellent thermal reliability within at least 200 melting/freezing cycles.

Microencapsulation is another approach to design very similar material like form stable PCM as the capsule confine the PCM and having gap inside can afford solid/liquid phase transition without damage [19]. The polymeric shell has a diameter in the range of 1-1000 μm. The phase change material resides in the core covered by a shell of natural or synthetic polymer as shown in figure 1.

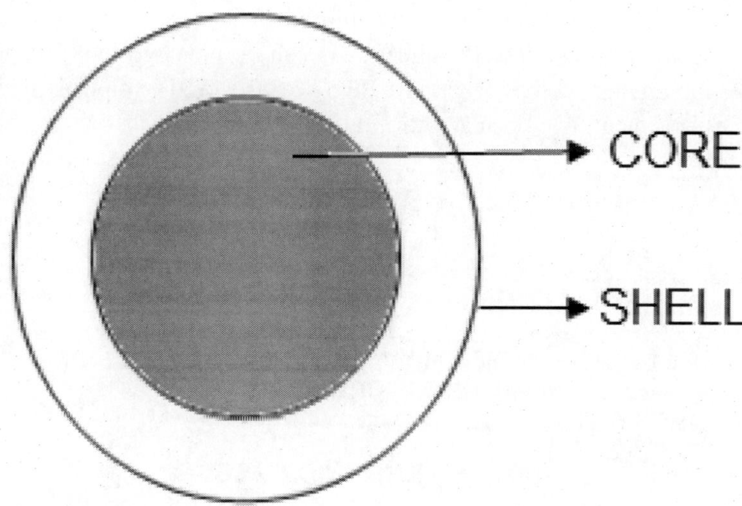

Figure 1. Representation of PCM microcapsule.

Apart from above two forms, others depend on reactivity of PCM material. Organic acids and polyethylene glycol are the two classes of material which have high latent heat of melting and excellent reactivity. The reactivity is a must for forming a chemical bond to ensure solid-solid phase change as it can be directly linked with cellulose cloth and others having reactive group present in the structure. Further, it can be incorporated in foam by reacting during foam formation. PEG being a versatile polymer having chemical reactivity, further discussion will be revolving around PEG.

Polyethylene glycol is soluble in many solvents including water. It has wide range of industrial applications. In fact the application as heat transfer fluid already exists for this material. The chemical reactivity of polyethylene glycols is mainly confined to the two terminal hydroxyl groups, which can be either esterified or etherified. It can react also with diisocyanate to form polyurethane.

There can be a range of such materials prepared via formation of chemical bond, such as:

- The copolymer poly (ethylene terephthalate)-PEG with varying phase change enthalpy [20].
- Rigid polymer cellulose diacetate (CDA) can be used as main chain with PEG chemically grafted as a functional branch chain [10]. However, the material is found unfit for melt processing.
- Another series of solid-solid PCMs can be prepared by synthesizing polyurethane by reacting PEG with MDI (diphenylmethane diisocyanate) and pentaerythritol [21].

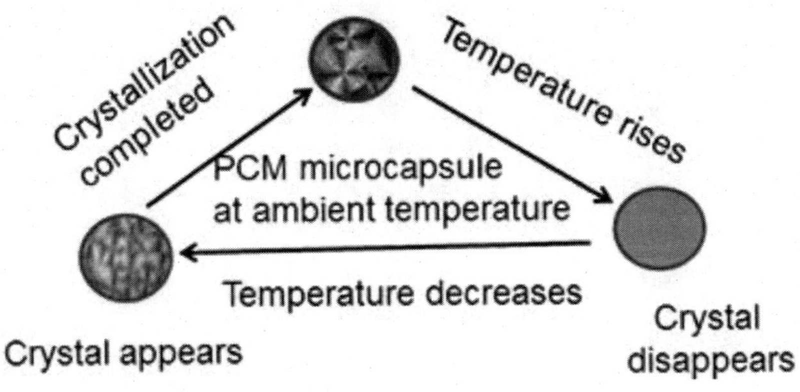

Figure 2. States of PEG during phase change in microcapsule.

Form Stable Polyethylene Glycol as PCM

For materials which flow on melting the encapsulation is the most convenient strategy and there are various options to realize the microcapsules filled with PCM. The changes in microcapsule filled with PCM during heating and cooling are shown in figure 2.

However, it is not free from drawback. One drawback of the encapsulation is that it has to be designed individually for each special application. Further, the low thermal conductivity of most PCMs is not conducive for applications where high power densities are required. The form stable one is another convenient strategy in which the conductivity can be tackled easily as the process is mostly physical in nature. For making form stable material another chemical or porous matrix is added to solid-liquid PCM system. Due to interaction via hydrogen bond, polar force, van der Waals force etc. the liquid phase is arrested and the blend behaves like stable solid-solid phase transition system. The hollow rayon and polypropylene fibers were impregnated with polyethylene glycols (having average molecular weights of 400, 600, 1,000 and 3,350) from 57% aqueous solution [22]. The thermal storage and release values of modified fibers are useful at temperatures as low as -8 to -48°C on cooling and as high as +42 to +77°C on heating with little variability. The heating and cooling performance are found to be distinct from untreated fiber as the treated fiber shows heat content around double as compared to the other. Further, even after 50 heating/cooling cycles the performance remains reproducible. In continuation with impregnation of PEG in hollow fiber the strategy was further strengthened by insolubilizing it with the help of crosslinker [23]. In addition to thermal properties, it imparts many additional attributes to the system, such as soil release, resistance to static charge, abrasion resistance, pilling resistance and water absorbency etc. Several cross-linking agents can be used such as, 1, 3-bis (hydroxymethyl)-4,5-dihydroxyimidazolidinone-2 etc. After removal of solvent the crosslinking occurs via ionic mechanism or under high energy radiation.

PEG/polymer Blend

Blend of PEG with cellulose was found to undergo dramatic transition when cellulose exceeded 5wt% [24]. While carrying out melting transition study, it is observed that there is no solid-liquid transition. It is only solid-solid transition even after exceeding melting temperature of PEG while exhibiting phase change enthalpy up to 100J/g. In a different study the cellulose and PEG blend was prepared by dissolution of former in dimethyl sulfoxide (DMSO) along with paraformaldehyde, adding latter at high temperature and mixing and stirring at room temperature [25] the weight ratio of PEG and cellulose is in the range of 95/5 to 50/50 (w/w). All the compositions exhibit solid-solid phase transition as no liquid comes out of blend even beyond melting temperature. Another report studied the application possibilities of PEG/cellulose blend for central processing unit (CPU) cooling application [26]. The PEG samples having varying molecular weight were blended with cellulose to make form stable PCM. Temperature gradient was analyzed for knowing about the location getting heated to maximum and minimum temperature respectively and PCM was applied accordingly. The phase-change behavior of PEG in composite with cellulose diacetate (CDA), prepared from the miscible solution, is found to be completely different from that of pure PEG when its fractional amount is less than 85% [27] . The PEG and CDA usually are miscible in acetone solution which makes the process simple. Even when the temperature is raised 40°C higher than the melting point of PEG, the solid–solid phase-change behaviour does not change. A series of biodegradable form stable blends of PEG were reported by preparing PEG/cellulose, PEG/agarose, and PEG/chitosan PCMs blends [28] . The PEG concentration was varied in the range of 60-80wt% which conformed to no leakage system at and above corresponding melting transitions. At melting point around 57°C the latent heats vary as the minimum of 84.63J/g, and maximum of 152.16J/g respectively are observed. The accelerated thermal cycling tests were applied successfully to ascertain thermal reliability and chemical stability.

PEG/Polyacrylate Blend

PMMA/PEG blends were prepared at different mass fractions of PEG (50, 60, 70, 80, and 90% w/w) using solution casting method and the exact form stable composition was checked from melting study [29]. The 70/30 (w/w) PEG/PMMA blend matched the characteristics of making form stable composition. The optical micrograph (Figure 3) indicates that PEG is uniformly distributed in PMMA matrix.

As expected, the melting and crystallization enthalpies are lower than pure PEG. In a parallel study the form stable PCM was realized by blending PEG with polymethyl methacrylate, acrylate copolymers and acrylate/acrylic acid copolymers respectively [30]. The PEG can be blended to a maximum of 80% w/w with any of the polymer/copolymer and no leakage of PEG is observed for 100 heating/cooling cycles. The phase behavior study indicates that the self- organized PEG is distributed in the matrix of acrylic polymers rather than having chain-like structures. The blends were planned for application in storing daytime solar energy to be used for space heating. Poly (acrylic acid) (PAA) or poly(ethylene-co-acrylic acid) (EcoA) were blended with different molecular weight of PEG (1000, 6000, and 10,000 g/mole) using solution blending procedure to make form stable PCM [31]. The acid group containing polymers form inter-polymer complexes (IPCs) and miscible and immiscible IPC–PEG blends. The blends below the bleeding out level are earmarked as potential materials for passive thermal energy storage applications.

A series of PEG having varying molecular weight were blended with a series of long chain fatty acids (capric, lauric, myristic, palmitic or stearic acid) to make thermal energy storage materials and investigation was done by using DSC [32]. The melt blending yielded form stable PCM. The blends show melting range from 30 to 72 °C and the latent heat of transition in the range of 168–208 J/g. Synergistic effect is observed with PEG 10000/stearic acid blend as the heat of transition is found to be approximately 15 and 35% higher than for pure stearic acid and PEG respectively. The hydrogen bonding helps in exhibiting synergistic effect via formation of more perfect crystalline lattice. The form-stable PCM/methacrylate employed in thermal management was designed by a completely different method. A monomer was polymerized in presence of PEG [33]. The composition prepared comprises PEG, polymethyl methacrylate (PMMA) and very important aluminum nitride (AlN), which serves as the thermal conductivity promoter. When the mass fraction of PEG is kept below 70%, the composite PCMs remain solid without leakage even above the melting point of PEG. The AlN additive enhances the heat transfer property of organic PCM which promotes the composition for thermal management of electronic device. Graphite nanoplatelets (GnPs), obtained by sonicating the expanded graphite, were used to harness 3 way properties of form stable PCM with PEG, such as thermal storage, thermal and electrical conductivity respectively [34]. The in situ polymerization method was conducted under ultrasonic irradiation in presence of GnPs (conductive fillers) and PEG, which were uniformly dispersed and embedded inside the network structure of PMMA. The strategy contributes to the well package and self-supporting properties of composite phase stable PCMs. XRD and FTIR studies indicate that the GnPs are physically combined with PEG/PMMA matrix and do not participate in the polymerization. The GnPs additives are able to effectively enhance the k and σ of organic phase stable PCM. At the mass ratio of GnP 8%, the k and σ of composite are enhanced to 9 times and 8 orders of magnitude respectively over that of pure PEG/PMMA matrix.

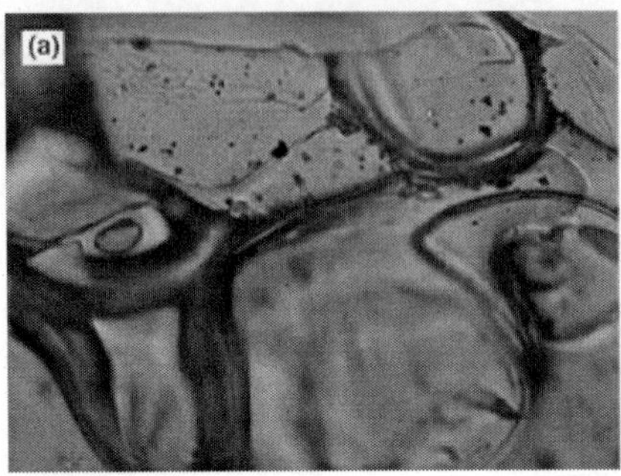

Figure 3. Optical Micrographs of (a) PMMA and (b) the form-stable PEG/PMMA blend.

Both thermally stable and form-stable PCM was prepared by blending PEG and polyamide 6 in formic acid with the use of polyether-ester-amide as compatibilizer [35]. The mass percentage of PEG went up to 80wt% while keeping the blend as form stable. The phase transition enthalpy of material increases with increasing the mass percentage and the molecular weight of PEG respectively.

PEG/Graphite Composite

Dual purpose form stable PCM is prepared by blending expanded graphite with PEG [36] as the porosity holds the PEG by preventing it to go to liquid state after melting and also enhance the thermal conductivity. The amount of PEG used was up to a maximum of 90wt% without any instance of leakage during the melting period. The latent heat of melting is around 161.2 J g^{-1}, which varies with composition. However, the melting point of 61.46°C almost remains constant in all the compositions. There is about 4 times enhancement of thermal conductivity of blend showing a value of 1.324 W m^{-1} K^{-1}. Graphene oxide (GO) sheets expectedly came out as excellent material for making composite with PEG by using melt blending process [37]. The PEG content goes up to 90 wt% in the composites to a form stable PCM while the heat storage capacity remains high at156.9 J g^{-1}, which is approximately 93.9% of the phase change enthalpy of pure PEG. Further, GO has much stronger influence on lowering of the phase change temperature of PEG as compared to other porous carbon materials (activated carbon and ordered mesoporous carbon). The unique thin layer structure of GO is the main reason for considering it as the promising heat storage candidate at mild temperature. Further enhancement of PEG content as high as 96% in form stable composition was achieved, which did not show any sign of leakage up to a temperature as high as 150°C [18]. The shape-stable PCM exhibits high heat storage capacity of 142.8 J g^{-1} and excellent thermal reliability even up to 200 melting/freezing cycles. The thermal properties of the PEG/GO composite PCMs with various GO contents are also investigated. A PEG/sulfonated graphene (SG) phase change composite was prepared by processing in aqueous medium to make form stable PCM with enhanced thermal performance [38].The addition of only 4 wt.% of SG to PEG results in four times increase in thermal conductivity with only slight decrease in the phase change enthalpy. The unique 2-dimensional morphology of graphene as well as strong interface affinity between PEG matrix and SG nanosheets are the reason for formation of efficient thermal conductive network within the PEG matrix. The production of PEG/SG composites can be taken up to pilot scale due to the facile fabricating process. The comprehensive composite PCMs using PEG and graphene nanomaterials can be made for higher performance [39]. Graphene oxide (GO) nanosheets were used as supporting materials to keep the shape of PEG stable during the phase change process and graphene nanoplatelets (GNP) were incorporated as conductive fillers to enhance the thermal conductivity and electrical conductivity. The PEG/GO/GNP composite PCM with only 2 wt% GO and 4 wt% GNP hybrid fillers exhibits excellent performance as the thermal energy storage density is very high at 98.2% of pure PEG and high thermal conductivity of 1.72 W/mK which is about 490% higher than that of pure PEG. The composition also exhibits excellent electrical conductivity of 2.5 S/m. Further, excellent thermal reliability is observed after thermal cycling.

PEG/AC Composite

Polyethylene glycol (PEG) and mesoporous active carbon (AC) were blended by impregnation method [40]. The crystallinity of PEG in the PCMs decreases with the increase in the AC content and the activation energy decreases with higher PEG weight percentages. The phase change properties of the PEG/AC PCMs are correlated with adsorption confinement of the PEG segments by the porous structure of AC. Further, the perfect PEG crystallization suffers from interference by AC. Three types of shape-stabilized PCMs were prepared by blending PEG with expanded graphite (EG), active carbon (AC) and ordered mesoporous carbon (CMK-5) as supporting materials having varied pore structures [41]. The PEG content of 70 wt% for AC and 90% for both EG and CMK-5 makes the form stable PCM. The phase change enthalpy and the PEG crystallinity decrease by forming composite in the order AC<CMK-5<EG while keeping PEG content fixed. It is observed that the micrometer size pores exhibit the same level of shape stabilization as the nanometer size pores, while minimizing the enthalpy loss due to the interactions between the pores and the PEG chains.

PEG/Silica Composite

Free flowing silica powders can be mixed with a phase change material to make form stable PCM [42]. The phase change materials can be chosen from many series including PEG of varying molecular weight. Up to 80 wt% PCM can be added to silica powders. The powder like mix can be used in tableware items, medical wraps, garments, quilts and blankets, and in cement composition. The preparation of PEG/silicon dioxide (SiO_2) composites as form-stable, solid-liquid PCM for thermal energy storage was carried out by using microporous SiO_2 [43] Even at much higher temperature than melting point of PEG, the shape remains stable as PEG remains adsorbed on microporous SiO_2. The favorable heat capacity and high thermal conductivity made it a candidate for solar energy and some other applications. Exactly similar work was published by the same author by explaining that when SiO_2 percentage goes beyond 15 wt% the composite remained solid even after crossing melting temperature [44]. The modulus values were also studied with dynamic mechanical analyzer (DMA). Mesoporus matrix such as activated carbon and silica molecular sieves were combined with high loading of PEG to make form stable PCM [45]. The significant features for PEG/AC PCMs are the capillary forces and surface areas, whereas, in PEG/silica PCMs capillary forces, surface areas, hydrogen bonding and surface polarities are more prominent. However, the porous characteristics and surface properties of the mesoporous stabilizers combination affect the crystallinity and phase change behavior of PEG. The outstanding properties such as largest latent heat, a relatively low melting point, the least supercooling and a higher heat storage efficiency are observed with 80wt% PEG PCM. A high conductivity form-stable phase change material was prepared by blending PEG, silica gel, and [β]-Aluminum nitride powder [46]. There is no change on the melting temperature with different composite ratios. The value of thermal conductivity changes from 0.3847 to 0.7661W/m/K with the increase of mass ratio of [β]-aluminum nitride from 5% to 30%. The heat storage and release rate of the composite PCMs are higher than that of pure PEG. The thermal conductivity enhancement of PEG/SiO_2 hybrid form-stable PCM was achieved by Cu doping via in-situ chemical reduction of $CuSO_4$ through ultrasound-assisted sol–gel process [16]. It is confirmed that Cu is in zero valence state and no new chemical bond exists among Cu, PEG6000 and SiO_2. The phase change enthalpy of $Cu/PEG/SiO_2$ PCM reaches up to 110 J/g, and the enhanced thermal conductivity is 0.414 W/m/K for 2.1 wt% Cu in PEG/SiO_2. The $Cu/PEG/SiO_2$ hybrid material exhibits excellent thermal stability and good form-stable performance

PEG/Nanoclay Composite

The first example of the direct intercalation of an organic polymer into the interlamellar spaces of kaolinite happened to be with PEG [47] . The PEG 3400 and PEG 1000 were intercalated into kaolinite by displacing dimethyl sulfoxide (DMSO) from the DMSO−kaolinite intercalate. As TGA/DSC analysis indicates, the complete decomposition of the organic component of the oxyethylene-based organokaolinites did not occur beyond 1000°C. However, the phase change study was not reported by the author. Having good thermal conductivity, this material can make an excellent PCM. A granular heat storage material was developed which was targeted for the direct contact with a heat transfer fluid [48]. The granules consist of a core of porous cordierite impregnated with PEG. Further, the coating on surface was fixed by reacting PEG with diisocyanate to make polyurethane. The increase of the accessible pore volume, caused by a large fraction of very small pores in the ceramic granules, leads to a considerable decrease of the useful melting enthalpy. The separation of the material in many small fractions decreases the crystallinity as well as the melting enthalpy of the material. A mass stability of 99.3% and about constant enthalpy over 72 melting/crystallization cycles are observed. The PEG/diatomite form stable composite was prepared by vacuum impregnation method [49]. The Scanning electron micrograph (SEM) indicates that the pores are well impregnated by PEG. The enthalpy value of the composite (Approx. 87 J/g) is found to be very high as compared to other form stable composites. Further, even after 1000 heating/cooling cycles there is no change in FTIR spectra indicating high stability. TGA data also confirms good thermal stability. Exfoliated graphite (EG) was added at various proportions to enhance the thermal conductivity for fast response of PCM. The conductivity is doubled at 10wt% of EG. Nanocomposite based form stable PCM was prepared in the form of low density polyethylene (LDPE)/ PEG composite along with nano-layered montmorillonite (MMT) [50].The PCMs were perpared by intercalation melt compounding of PEG with MMT followed by melt blending with (LDPE) by using polyethylene grafted maleic anhydride (LDPE-g-MAH) as compatibilizer in a twin-screw extruder. The phase transition enthalpy shifted from 20.58 J/g to 88.76 J/g and also the phase transition temperature shifting from 47.5 °C to 63.6 °C.

In another nanoclay incorporated PCM, PEG and montmorillonite nanoclay were blended in epoxy matrix to prepare form stable PCM [51]. A special performance evaluation test was designed for evaluating the top surface temperature of prepared samples under back surface heating. The increasing PCM content improves the thermal protection performance. However, lower thermal diffusivity is found for the sample containing 60 wt% PEG, with a 31% decrease in top surface temperature. These results show that increase of top surface temperature of samples containing PCM is very slow when compared with the neat epoxy sample. The heat protection performances of nanocomposite blends having 5 and 7wt% of clay in PEG have been improved by about 10% as compared to epoxy /PEG60 blend.

PCM Derived from Chemically Reacted PEG

Chemically reacted PEG remains always in solid state so long as the other part of the polymer remains solid. This strategy is always considered as straightforward in nature and other ingredients can be manipulated as required for any application.

PEG Reacted with Cellulose and Its Derivative

There was study on individual behaviour of PCM based on both physically blended and chemically attached PEG with CDA [52]. It is observed that for a chemically bonded material, the bonding strength is strong enough to ensure that no PEG is detached at high temperature, whereas, in blends under identical condition, the amorphous PEG migrates freely within the CDA framework and bleeds out of the blend matrix through a diffusion mechanism. PEG grafted CDA makes chemically modified PCM as PEG is hooked to free hydroxyl group of CDA by first reacting with diisocyanate and later with CDA [10]. With PEG hooked as side chain the free mobility and reorganization to crystalline structure is impeded and rapid decrease of enthalpy is observed. Further, the crystal structure being imperfect, the melting transition temperature reduces. In fact in one case it went down to about 30°C. Phase change enthalpy increases rapidly with the PEG molecular weight reaching a maximum at molecular weight of 10000 followed by slight decrease. The cellulose-g-PEG copolymers were synthesized with different graft densities from MTritylcellulose [53]. As expected, the spherulite sizes of the copolymers are smaller than that in pure PEG. In PEG 2000 grafted sample the melting enthalpy decreases rapidly from 200 to 66 J/g as the PEG content decreases from 100 to 50wt% and the corresponding melting point decreases from 55 to 45°C respectively. The thermal stability of grafted PEG is also very good. The nano-crystalline cellulose/PEG based PCM was synthesized by grafting PEG as branch chain to cellulose main chain [54]. These PCMs have exhibited enthalpies up to 103.8 J/g. A series of cellulose graft PEG (cellulose-*graft*-PEG) copolymers as PCMs were synthesized in ionic liquid by using the coupling reactions between cellulose hydroxyl and PEG hydroxyl groups using 4, 4-diphenylmethane diisocyanate [55]. The phase change temperature is in the range of 40–60 °C. The copolymers are solid–solid PCMs with good thermal stability. For enhancing the thermal conductivity, the expanded graphite (EG) in various proportions (2, 5 and 10 wt%) was added to the PCMs.

PEG Reacted with Non-Cellulosic Polymers

The PEG10000/poly (glycidyl methacrylate) (PGMA) crosslinked copolymer as PCM was synthesized by reaction of end-carboxyl groups in carboxyl modified PEG and epoxy groups in PGMA respectively [56]. The WAXD patterns and POM images show that the crystalline form of the copolymer remains similar to that of pure PEG. The phase change temperature is found in the range of 25–60°C, and heat storage capacity around 70 J/g. Environment friendly PCM was obtained by copolymerizing poly(hydro-xybutyrate-*co*-hydroxyvalerate) and PEG (PHBV/PEG) through free-radical solution polymerization using PHBV diacrylate (PHBVDA) and polyethylene glycol diacrylate (PEGDA) as macromers [57]. Polarized optical micrograph (POM) of the copolymer exhibits the composite spherulites, with the interior part displaying the extinction rings of PHBV banded spherulites, the outside remains irregular, with clearly elongated radial growth indicative of PEG crystallization. The copolymers have excellent heat resistant properties. The melamine/formaldehyde/PEG (MFPEG) crosslinked copolymers, which can function as PCM, were synthesized using standard protocol of amine–aldehyde condensation reaction and aldolization respectively [58]. The MFPEG synthesized with PEG6000 has a good crystallization property. The melting enthalpy of MFPEG is less than that of pristine PEG for obvious reason as the PEG crystallization is restricted due to linking at both ends. The maximum latent heat enthalpies of heating and cooling cycle are found to be 109.4 and 103.9 J/g respectively. The MFPEGs are stable up to 340–455°C in nitrogen atmosphere. A PCM copolymer named poly (decaglycerol-co-ethylene glycol)

[P(DG-co-EG)] was prepared by free-radical solution polymerization from two macromonomers, PEG (4000) acrylate (PEGA) and polydecaglycerol acrylate (PDGA) respectively [59]. During transition from PEG to PEG acrylate to PEG in polymer chain the crystallization ability of PEG molecular chain reduces due to hindrance from attachment in the polymer architecture and the result is the incomplete crystallization of PEG chain, which forms micro-crystals. The melting and crystallization enthalpies of P(DG-co-EG) are 163.51 J/g and 141.31 J/g, respectively and the respective melting and crystallization temperatures are 51°C and 42°C respectively. The molecular chains of PEG reside as branched-chain of copolymer and also its presence in comb networks of copolymer acts in the direction of lowering of the crystallization integrity and enthalpy. The thermal stability of the copolymer is quite good. Its degradation occurs in multiphase mode in TGA measurement. The small weight loss during 210–240°C is assigned to impurity, which is followed by another weight loss during 310–330°C due to degradation of ester bond and at 330–420°C due to degradation of PEG. Finally, the degradation at 420°C is due to polymer backbone. The PEG was incorporated as PCM fragment in poly (PEG methyl ether methacrylate) via the bulk polymerization of polyethylene glycol methyl ether methacrylate [60]. The poly (PEG methyl ether methacrylate) shows good crystallization ability due to the flexibility of long polyether side chain. The facile synthesis, solid–solid transition at suitable transition temperature, high transition enthalpy, and good thermal stability make this material a potential material for application in energy storage.

PEG Reacted to Form Polyurethane (PU)

The polyurethane block copolymer PCM was designed by reacting high molecule weight PEG as soft segment, 4,4′-diphenylmethane diisocyanate (MDI) and 1,4-butanediol (BDO) (as a chain extender) by a two-step process [61]. The PCM exhibits typical solid–solid phase transition properties, e.g., suitable transition temperature, high transition enthalpy and good thermal stability. The urethane linkage, a hard segment, serving as "physical cross-links," restricts the molecular chain of the soft segments from free movement at high temperature. Polyurethane PCMs, with various molecular weights PEG as soft segments and MDI-BDO forming hard segments were synthesized by two-step solution polymerization method [62]. The crystallizability of soft segment is found depressed by the introduction of hard segments in the polyurethane architecture. Further, the soft segments are not able to crystallize until its average molecular weight is above 2000. The resultant PU shows phase transition reversibility and stability throughout the cycling process. A crosslinked PCM was synthesized via two-step condensation reaction of high molecule weight PEG (MW: 10000) with pentaerythritol (PE) and MDI [22]. The cross-linked PCM shows typical solid–solid phase transition property having phase change enthalpy and crystallinity up to 152.97 kJ/kg and 81.76%, respectively. Extensive study on PEG based polyurethane was done by Meng et. al.[63]. PEG was introduced as the soft segment in shape memory thermoplastic polyurethane (PU). The well-formed phase separated structure in the PEG-based polyurethane (PEGPU) shows combined properties of PCM and shape memory polymer. Optical micrograph indicates that spherulites in PEGPU are smaller than and not as perfect as those in pure PEG. Due to the hydrogen bonded hard segment phase serving as "physical cross-link" the free movement of the soft segments is restricted at temperature above the PEG phase melting transition. The PEGPU has latent heat storage capacity higher than 100 J/g and melting point at 47°C. Beyond 50°C, the material shows a plateau in storage modulus (E′) plot at about 40MPa, which is due to the hydrogen bonded hard segment phase. When the temperature goes above 150°C, another mild decrease in E′ is observed, indicating the disappearance of hydrogen bonded crosslinks [64]. The melt flow index results indicate that the material has good melt-processing ability. Bis(1,3-dihydroxypropan-2-yl) 4,4′-methylenebis (4,1-phenylene)dicarbamate, a tetrahydroxy compound, was designed and synthesized as PCM [65]. The thermoplastic polyurethane solid–solid PCM (TPUPCM) was prepared by using PEG as soft segment and 4,4′-diphenylmethane diisocyanate and tetrahydroxy compound made up of multi-benzene ring structure as hard segments. The resultant PCM possesses excellent phase-change property and an applicable temperature range. The melting and the crystallization phase-change enthalpy are 137.4 J/g, and 127.6 J/g respectively. The decomposition of TPUPCM starts at 323.5°C and has good mechanical properties. The solid–solid phase-change material is dissolvable, meltable and can be processed directly. Hyperbranched polyurethane copolymer PCM (HB-PUPCM) using hyperbranched polyester as chain extender was prepared via a two-step process [66]. The phase transition behavior and morphology of the films indicate that the polymer is a good solid–solid phase change heat storage material. In DMA study, the plateau is observed, which is similar to that of other PU results. The tanδ curve does not come back to the base line in plateau region indicating damping over extended temperature range.

PCM Fiber with PEG

Energy saving has great role in the field of energy storage. PCMs being one energy storage material must have some role to play. They can produce high density of energy storage during phase change. The stability must be ensured to make it commercially viable. However, there is a risk of material leakage during their phase change. Electrospinning is one of the methods which can make thermo-regulating nanofibers of leakage free form-stable PCMs. Many approaches deal with various aspects of the nanofiber.

Electrospinning of PEG /cellulose acetate (PEG/CA) composite was used to prepare ultrafine PCM fibers [67]. The morphology of the fiber is cylindrical and has a smooth external surface. The distribution of PEG is found to be on the surface as well as within the core of the fibers. The fibers impart balanced thermal storage and release properties for their thermo-regulating function and even after 100 heating–cooling cycles the thermal properties are reproducible. The same group studied further the effect of molecular weight of PEG on the behaviour of the form-stable polymer-matrix phase change composite electrospun ultrafine fibers for thermal energy storage [68]. The cellulose acetate and five different molecular weight (M_n) grades of PEG were used for the study. The composite fibers are smooth and cylindrical in shape, having average diameters in the range of 1000--1750 nm. Further, the diameter increases with M_n of PEG. The composite fibers exhibit varied thermal storage and release properties according to the temperature ranges with the variation of M_n of PEG. Excellent thermal stability and reliability are observed even after 100 heating-cooling thermal cycles. Another study was done by the same group to improve their water-resistant ability and thermal stability for potential thermal storage applications [69]. The electrospun PEG/CA fibers were crosslinked by using toluene-2, 4-diisocyanate (TDI) via urethane linkage. The stability study by immersing in deionized water confirmed that the properties of crosslinked fiber are preserved even after 24 h. As expected the crosslinking improves the thermal stability, while the phase change enthalpy decreases substantially. The study was extended further by increasing the PEG content and finding the effect on the morphology, crystalline properties, phase change behavior and tensile properties of the composite fibers [70]. As observed in SEM micrograph the maximum PEG content in the fibers could reach up to 70wt%. However, the morphology and average diameter of the composite fibers vary with PEG content. The latent heat of the composite fiber increases with the PEG content in the fiber.

The PET-PEG copolymer fibers were successfully prepared by melt spinning [20]. The copolymer has solid-solid phase change characteristics at 10–60°C without any liquid substance separation. In comparison, the PET/PEG blends lose their phase-change characteristics since the PEG of the blends melts and leaks under high temperature. By controlling the molecular weight and proportion of PEG added, the phase-change temperature range and the enthalpy can be adjusted.

PEG 6000 together with nylon 6,6 as supporting material was electrospun in different blend ratios [71]. For different blend ratios, the average nanofiber diameters observed are in the range of 178 to 507 nm. The composite samples have high heat capacity and good energy saving properties. To make the system more stable, the as-spun composite nanofibers were successfully changed to a crosslinked structure by using a suitable cross-linking agent.

Nonwoven mats were fabricated from a blend of polyvinylidene fluoride (PVDF) and PEG 1000 by carrying out electrospinning [72]. The PVDF acts as supporting material to afford a mechanically strong form-stable structure. However, fumed silica was incorporated in the electrospun PVDF/PEG 1000 nanofibers to prevent the leakage of molten PEG 1000 during the solid–liquid phase change. Further, the addition of fumed silica also helps in improving the mechanical strength of the PVDF/PEG 1000 nonwoven mats. There are no significant changes in the melting temperature and heat of fusion of PEG 1000 in the composite nanofiber after 100 heating/cooling cycles. PEG/ PVDF composite core/shell nanofibers were fabricated using coaxial electrospinning [73]. In the process, the melted PEG and PVDF were coaxially electrospun through a double spinneret as a core layer and as a shell layer respectively. The PEG remains as core layer in the composite nanofibers with PVDF shell to prevent its leakage and reduce the effect of the external environment during use. The core/shell structure is confirmed by water contact angle (WCA) measurements. The composite nanofiber has good thermal stability and energy storage capacity. Three different molecular weights (1000, 2000 and 4000) were used as the core material to prepare composite nanofibers with different melting/crystallization temperature ranges and thermal storage capacities. The WCA value of 106° of PEG4000/PVDF core/shell nanofibers is similar to that of PVDF nanofibers confirming that the PEG 4000 is completely covered by PVDF at the highest core feed rate of 0.210 mL/h. Further, at this core feed rate, the core/shell nanofibers has the largest content of PEG in the core up to 42.5 wt% with a latent heat of 68 J/g and a melting temperature of 62.8 °C. These shape-stabilized core/shell nanofibers show good thermal reliability and high tensile strength as well.

Similarly, ultrafine phase change fibers (PCFs) with core–sheath structure based on PEG /cellulose acetate (PEG/CA) blends were fabricated by employing coaxial electrospinning method [74]. SEM and TEM microstructures suggest that cylindrical and smooth phase change fibers are obtained and the structure mimics PEG encapsulation with CA shell. The composite fiber exhibits balanced and reversible phase change behavior, and phase transition enthalpies, which increase with the PEG content in the fibers. The phase transition temperatures of the fibers are exactly similar to that of pure PEG. The mechanical property study indicates ultimate strength and ultimate strain of the composite fibers are lower than those of CA fibers, which further decrease with increase of PEG content.

The poly (N-hydroxymethyl acrylamide)/PEG (PNHMPA/PEG) IPN gel was prepared as a form-stable composite PCM [75]. The IPN was used to prepare PCM fibers by incorporating in polypropylene (PP) matrix and utilizing melt spinning process. The PNHMPA/PEG IPN has porous structure as observed in SEM micrograph. Further, no phase separation is noticed in multicomponent matrix and the PCM IPN is distributed evenly in PP. However, the tensile strength of PP/IPN fibers is slightly lower than that of pure PP fiber. The melting temperature and the latent heat of the composite are 55°C and 82 J/g respectively and the thermal stability is also very good. Further, the damping performance of PCM/IPN composite fibers is very good due to presence of IPN.

PCM Textile with PEG

The most comfortable human skin temperature is 33.4°C. Once the fluctuation in the temperature of the skin exceeds ± 4.5°C, the human body will feel discomfort [76]. The new textile insulation materials with heat storage attributes have become important because they regulate temperature according to the environment and body temperature. These insulation textiles are called thermo-regulated textiles, and contain PCMs [77]. As a common technique of integration of PCM with textile, the microcapsule technology comes handy. Melamine–formaldehyde microcapsules containing eicosane was prepared by *in situ* polymerization [78]. The microcapsules were added to polyester knit fabrics by a conventional pad–dry–cure process to develop thermoregulating textile materials. PEG 600 has been used as a phase change material (PCM) to make a thermo-regulating fabric [79]. Urea formaldehyde prepolymer and PEG 600, with other ingredients at 40°C, made oil-in-water system and finally PEG was engulfed by the resin by forming resin microcapsule shell. Microcapsules were coated on a 100% cotton fabric. The fabric was dip coated by immersing in a bath containing PU and microcapsules (20:80, wt/wt) while the former acted as binder. The fabric was padded and cured at 80°C and tested for performance. The fabric specimen was fixed on the sweating hot plate in such a way that the coated side faced the hot plate. The heat released from the fabric was monitored by a sensor placed 7 mm above the fabric. The sensor above the treated fabric detected 20% less heat than that of the untreated fabric. A similar work of applying pad-dry-cure method for adhering PCM on textile was reported by Karthikeyan et. al. [80]. The nanoencapsulated PCMs containing PEG as the core material and urea formaldehyde as the shell material were prepared by using in situ polymerization method. The pad-dry-cure method was applied in different ratios of nanocapsules to binder. The nanocapsules are of regular spherical shape having size around 141 nm. The treated fabrics show latent heat storage capacity of 0.198, 0.213 and 0.219 J/g at different nanocapsules/binder ratio. The treated fabric with a nanocapsules/binder ratio of 3:3 exhibits good thermal stability.

A little variation of approach was made by encapsulating carbowax along with PEG 8000 or calcium chloride [81]. The shell is an elastomeric condensation polymer, such as a polyurethane-urea. No significant lowering of the latent heat of fusion of the solid PCM is observed on encapsulation with a condensation polymeric shell. During usage the encapsulated phase change material picks up sensible heat after phase change. The encapsulated PCM shows good integrity and the elastomeric polymeric shell is able to accommodate the volume change during phase change.

Considering the dislodgement of microcapsule attached to textile garments during repeated washing, ironing etc., direct attachment via chemical linkage merits more attention as PCM molecule/polymer becomes integral part of the garment. Attachment of PEG via chemical linkage formation (polyurethane) was proposed [82]. The diisocyanate molecule can react at one end with one hydroxyl of PEG and the other end with hydroxyl of cellulose. The reaction scheme is presented in Scheme 2

By grafting this way, the PEG remains solid during phase change as it is chemically anchored to textile. The melting is observed at 60°C for PEG 8000 and corresponding enthalpy is around 55-59 J/g. Initial degradation temperature (IDT) is around 350°C. Tear strength of cotton shows mild decrease. The water and perspiration fastness and rubbing fastness remain same like that of unmodified cotton. Interestingly, 13% increase in abrasion resistance is observed. Most of the inherent properties of the fabric remain unchanged while phase change attribute is imparted by direct treatment.

PEG As PCM in Concrete and High Performance Composite

For thermal energy storage in buildings the PCMs have been considered since the period before 1980. With the facile incorporation of PCM in gypsum board, plaster, concrete or other wall covering material, thermal energy storage is becoming a reality as it can be part of the building structure even for light weight buildings [83]. The development and testing were conducted for prototypes of PCM wallboard and PCM concrete systems to enhance the thermal energy storage (TES) capacity of standard gypsum wallboard and concrete blocks, with particular interest in peak load shifting and solar energy utilization.

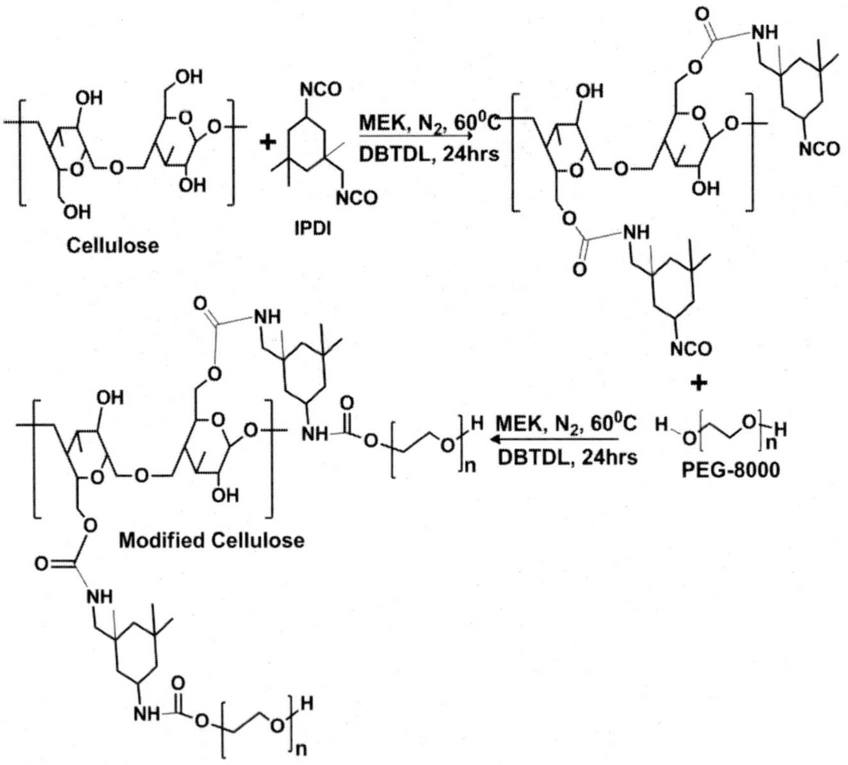

Scheme 2. Reaction scheme for synthesis of PEG grafted cellulose.

Solar energy applications need an efficient thermal storage system/ material as the successful application of solar energy depends on the method of energy storage used. The amount of the energy involved can be demonstrated by comparing the sensible heat capacity of concrete (1.0 kJ/kg K) with the latent heat of a PCM, such as calcium chloride hexahydrate $CaCl_2$, $6H_2O$ (193 kJ/kg). This indicates that any energy storage systems incorporating PCM will comprise significantly smaller volumes when compared to other materials storing only sensible heat [84].

The PEG was used as PCM in preparing PEG/cement composites [85]. The thermal properties and thermal stability were investigated in detail. The SEM micrograph shows well dispersed PEG in the porous network of the cement.

The influence of PEG as PCM on the performance and thermal stability of common asphalt and AC-20 asphalt concrete was studied [86]. It is observed that with the increase of PCM, the viscosity of asphalt composite increases while the ductility and temperature susceptibility show decline. When 30% PEG is mixed with asphalt, the modified asphalt can sustain 60°C for five hours. Further, the AC-20 asphalt/PEG mixture exhibits excellent high-temperature stability and the dynamic stability is twice as compared to common AC-20 asphalt mixture. The pavement temperature is found to be 9°C lower than the summer day temperature of 39°C. However, at low temperature, no effect is noticed.

It is well known that PCM can be used to reduce the temperature rise of a large concrete section during (semi)adiabatic curing, so that thermal cracking is minimized. Also, a PCM can reduce the number or intensity of freeze/thaw cycles experienced by a bridge deck or other concrete structure exposed to a winter environment. Both the applications were explored experimentally as well as by modelling [87]. It is observed that when PEG is impregnated into the porous lightweight aggregates (LWA), this PCM is found to substantially retard cement hydration. Thus, it appears that for successful utilization in fresh concrete, the PEG PCMs need to be encapsulated by suitable shell material.

In case of freezing, the pore solution in cementitious bodies expands and can creates stresses causing damage. The reduction in the frequency of freeze/thaw cycles for a structure will enhance the structure's service life. This fact necessitates incorporation of PCMs so that the number of freeze/thaw cycles experienced by bridge decks reduces, which has been investigated by modeling, mechanical testing, calorimetry, and X-ray microtomography [88]. The model can identify the geographical regions where the service life can be extended by one year by using PCM. Due to various reasons, the PCM incorporation usually reduces strength by varying amounts.

When PCM material is used, it contributes to the energy efficiency of buildings by reducing the peaks in the daily temperature cycles. The study was executed in one part of India, where temperature remains high throughout the year. The study was aimed at investigating the potential of using PCM for thermal heat storage in building facades in hot humid climates [89]. The PEG E600 with melting points 25 & 31°C respectively, was used as experimental material for study & analysis based on the weather data & the easily available material in the market. The PEG (PEG: E600) is known as chemically stable with high heat of fusion, safe and recyclable. Glass test tube was filled with E600 and inserted inside the hollow bricks of scale down model house (1 Cu. Ft.) built up with hollow bricks and temperature recorded between 9.00 am to 3.30 pm with the building kept in East to West direction. (Figure 3)

Up to 12.30 pm the temperature inside the building model was below the outdoor ambient temperature spanning four and half hour.

Two different lightweight aggregates (an expanded clay and a naturally porous pumice) and four different PCMs (PT4, PCM6, PEG400, and PEG600) have been used for incorporation in to cementous compositions [90]. The abilities of the aggregates to hold PCMs have been investigated, along with the effects of LWA/PCM incorporation on early age properties, strength, and thermal properties of mortars. The data is used to update a previously developed service life prediction model [91]. The earlier model is refined, while more experimental data and fewer assumptions are adopted. The most important property governing the effectiveness of PCM at reducing freeze/thaw cycling is thermal diffusivity, or in other words, how quickly the thermal equilibrium is established with the environment. The model results indicate that a low rate of heat transfer through the material is more effective. Considering everything, roughly a conservative dose of 50 kg/m^3 of PCM would enhance bridge deck service life up to a minimum of one year.

A number of PU rigid foams designed with three types of PEGs and used as insulation materials with enhanced thermal capacity were studied for finding the suitability for various applications such as layer of floor and ceiling coverings in constructions, insulations in controlled temperature transportation packaging, inner coverings of automobile seats, etc. [92]. Two-layer concrete–PU foam system was designed and studied under laboratory conditions for the insulation performances via using a computer-aided thermal measurement setup which was sensitive to the simulated environmental temperature changes. The PUI, including 44% PEG 600, exhibits efficient thermal regulation under moderate ambient temperature conditions, whereas PUII (49% PEG 1000) is suitable for controlling both mild and hot surroundings. PUIII, containing 53% PEG 1500, shows suitable heat storage and thermal stability characteristics. PUIV, containing 38% PEG 600/PEG 1000/PEG 1500, also confirms good thermal and durability characteristics. The blend of three PEGs is found to be suitable for preventing discontinuous thermal condition during external temperature fluctuation. PU foams containing PEGs are expected to be leak proof, which make it promising for industrial applications.

A composition comprising a polymeric material and PEG having molecular weight greater than 400 or end-capped PEG as a phase change material is useful for realizing products like molded and/or coated materials such as flooring, tiles, wall panels, paints and similar items [93]. The composition exhibits melting enthalpy of about 30cal/g.

Poly (N-hydroxymethyl acrylamide)/PEG (PNHMPA/PEG) interpenetrating polymer network (IPN) gel was synthesized as a form-stable phase change thermal filler [94]. The synthesized IPN was disbursed evenly in a phase change thermal conversion coating. The crystalline nature of IPN as observed in polarizing microscope indicates that IPN can function as heat storage material in the coating. In fact, the DSC data supports this as the conversion coating has a good energy storage capability at required phase transition temperature. The thermal stability of the coating is also very good.

Figure 4. Building model setup: PEG E600 integrated hollow brick walls covered by RCC Slab.

REFERENCES

[1]   BP Energy Outlook 2035, 15 Jan, 2014, www.bp.com/ energyoutlook.

[2]   Thermal Energy Storage: Technology Brief, IEA-ETSAP and IRENA©
       Technology Brief E17 – January 2013; www.etsap.org – www.irena.org.

[3]   Energy Conservation through Energy Storage (ECES) Programme,
       International Energy Agency, Brochure: http:// www.iea-eces.org/ fi
       les/090525_broschuere_eces.pdf.

[4]   Demirbas, F. *Energ. Source. Part B.*, 2006, 1, 85-95.

[5]   Mehling, H; Cabeza, LF. Heat and cold storage with PCM: An up to date
       introduction into basics and applications; *Solid-liquid phase change
       materials* and 978-3-540-68557-9; Springer: Heidelberg, Berlin, 2008,
       Vol. 1, 11-55.

[6]   Lane, GA. Solar heat storage: latent heat materials; Background and
       scientific principles; CRC Press, Inc.; Florida, 1983, vol. I.

[7]   Sarier, D; Chuah, TG; Salmiah, A; Choong, TSY; Saari, M. *Int. J. Green
       Energ.*, 2004, 1, 495–513.

[8]   Sarier, N; Onder, E. *Thermochim. Acta*, 2012, 540, 7–60.

[9]   Kenisarin, MM; Kenisarina, KM. *Renew. Sust. Energ. Rev.*, 2012, 16,
       1999–2040.

[10]   Jiang, Y; Ding, E. *Polymer*, 2002, 43 (1), 117-122.

[11]   Zhou, X; Xiao, H; Feng, J; Zhang, C; Jiang, Y. *Compos. Sci. Technol.*,
       2009, 69, 1246–1249.

[12]   Zhou, X; Xiao, H; Feng, J; Zhang, C; Jiang, Y. *Chem. Eng. Res. Des.*,
       2010, 88, 1013–1017.

[13]   Karaipekli, A; Sarı, A. *Renew. Energ.*, 2008, 33, 2599–2605.

[14]   Sarı, A; Karaipekli, A. *Mater. Chem. Phys.*, 2008, 109, 459–464.

[15]   Sarier, N; Onder, E; Ozay, S; Ozkilic, Y. *Thermochim. Acta*, 2011, 524,
       39–46.

[16]   Tang, B; Qiu, M; Zhang, S. *Sol. Energ. Mat. Sol. C.*, 2012, 105, 242–
       248.

[17]   Fan, L-W; Fang, X; Wang, X; Zeng, Y; Xiao, Y-Q; Yu, Z-T; Xu, X; Hu,
       Y-C; Cen, K-F. *Appl. Energ.*, 2013,110, 163–172.

[18]   Qi, G-Q; Liang, C-L; Bao, R-Y; Liu, Z-Y; Yang, W; Xie, B-H; Yang,
       M-B. *Sol. Energ. Mat. Sol. C.*, 2014, 123,171–177.

[19]   Ren, YJ; Ruckman, JEJ. *Cloth. Sci. Tech.*, 2004, 16 (3), 335-347.

[20]   Hu, et al. *J Macromol. Sci., Part B: Phys.*, 2006, 45, 615-621.

[21]   Li, WD; Ding, EY. *Sol. Energ. Mat. Sol. C.*, 2007, 91, 764-768.

[22]   Vigo, TL; Frost, CMJ. *Ind.Text.*, 1983, 12 (4), 243-254.

[23] Vigo, TL; Zimmerman, CM; Bruno, JS; Danna, GF. *United State Patent No.*, 4,908, 238, Mar. 13, 1990.

[24] Guo, YQ; Ding, EY; Gu, LZ; Liang, XH. *J. Macromol. Sci. Part B Phys.*, 1995, 34 (3), 239-248.

[25] Liang, X-H; Guo , Y-Q; Gu , L-Z; Ding, E-Y. *Macromolecules*, 1995, 28 (19), 6551–6555.

[26] Wojda, M; Wójcik, TM. *MATEC Web of Conferences*, 2014, 18, 03009, EDP Sciences.

[27] Guo, Y; Tong, Z; Chen, M; Liang, XJ. *Appl. Polym. Sci.*, 2003, 88, 652–658.

[28] Senturk, SB; Kahraman, D; Alkan, C; Gokse, I. *Carbohyd. Polym.*, 2011, 84 (1), 141-144.

[29] Sarı, A; Alkan, C; Karaipekli, A; Uzun, O. *J. Appl. Polym. Sci.*, 2010, 116, 929–933.

[30] Alkan, C; Sari, A; Uzun, O. *AIChE J.*, 2006, 52, 3310–3314.

[31] Alkan, C; Gunther, E; Hiebler, S; Himpel, M. *Energ. Convers. Manage.*, 2012, 64, 364–370.

[32] Pielichowski, K; Flejtuch, K. *Macromol. Mater. Engg.*, 2003, 288 (3), 259–264.

[33] Zhang, L; Zhu, J; Zhou, W; Wang, J; Wang, Y. *Thermochim. Acta.*, 2011, 524 (1-2), 128-134.

[34] Zhang, L; Zhu, J; Zhou, W; Wang, J; Wang Y. *Energy*, 2012, 39 (1), 294–302.

[35] Guo, J; Li, N; Xu, D-Z; Wang, L-Y. *Polym. Mater. Sci. Engg.* 2009, 5 (16), 1-164.

[36] Wang, W; Yang, X; Fang, Y; Ding, J; Yan, J. *Appl. Energ.*, 2009, 86 (9), 1479-1483.

[37] Wang, C; Feng, L; Yang, H; Xin, G; Li, W; Zheng, J; Tian, W; Li, X. *Phys. Chem. Chem. Phys.*, 2012, 14, 13233-13238.

[38] Li, H; Jiang, M; Li, Q; Li, D; Chen, Z; Hu, W; Huang, J; Xu, X; Dong, L; Xie, H; Xiong, C. *Energy Convers. Manag.*, 2013, 75, 481-487.

[39] Qi, G-Q; Yang, J; Bao, R-Y; Liu, Z-Y; Yang, W; Xie, B-H; Yang, M-B. *Carbon*, 2015, 88, 196–205.

[40] Feng, L; Zheng, J; Yang, H; Guo, Y; Li, W; Li, X. *Sol. Energ. Mat. Sol. C.*, 2011, 95(2), 644–650.

[41] Wang, C; Feng, L; Li, W; Zheng, J; Tian, W; Li, X. *Sol. Energ. Mat. Sol. C.*, 2012,105, 21-26.

[42] Salyer, IO. 1993, United State Patent Number 5cao21149, May 18, 1993.

[43] Wang, W; Yang, X; Fang, Y; Ding, J; Yang, J. *CIESE J.*, 2007, 58(10), 2668-2668.

[44] Wang, W; Yang, X; Fang, Y; Ding, J. *Appl. Energ.*, 2009, 86 (2), 170-174.

[45] Feng, L; Zhao, W; Zheng, J; Frisco, S; Song, P; Li, X. *Sol. Energ. Mat. Sol. C.*, 2011, 95 (12), 3550–3556.

[46] Wang, W; Yang, X; Fang, Y; Ding, J; Yang, J. *Appl. Energ.*, 2009, 86 (7-8), 1196-1200.

[47] Tunney, JJ; Detellier, C. *Chem. Mater.* 1996, 8 (4), 927-935.

[48] Waschull, J; Guhlemann, T. *2nd International Renewable Energy Storage Conference (IRES II)*, 19-21 November. 2007, Bonn, Germany.

[49] Karaman, S; Karaipekli, A; Sarı, A; Bicer, A. *Sol. Energ. Mat. Sol. C.*, 2011, 95, 1647–1653.

[50] Zhou, X; Xu, W. *Polym. Mater. Sci. Eng.*, 2011, 3, 140-142

[51] Bahramian, AR; Ahmadi, LS; Kokabi, M. *Iran.Polym. J.*, 2014, 23 (3), 163-169.

[52] Ding, E-Y; Jiang, Y; Li, G-K. *J. Macromol. Sci. Part B: Phys.*, 2001, 40 (6), 1053-1068.

[53] Li, Y; Liu, R; Huang, Y. *J Appl. Polym. Sci.*, 2008, 110, 1797–1803.

[54] Yuan, X-P; Ding, E-Y. *Chinese Chem. Lett.*, 2006, 17, 1129-1132.

[55] Li, Y; Tang, M; Liu, R; Huang Y. *Sol. Energ. Mat. Sol. C.*, 2009, 93 (8), 1321-1328.

[56] Chen, C; Liu, W; Yang, H; Zhao, Y; Liu, S. *Sol. Energy*, 2011, 85 (11), 2679-2685.

[57] Xiang, H; Wang, S; Wang, R; Zhou, Z; Peng, C; Zhu, M. *Sci. China Chem.*, 2013, 56 (6), 716–723.

[58] Li, Y; Wang, S; Liu, H; Meng, F; Ma, H; Zheng, W. *Sol. Energ. Mat. Sol. C.*, 2014,127, 92–97.

[59] Guo, J; Xiang, H; Wang, Q; Hu, C; Zhu, M; Li, L. *Energ. Buildings*, 2012, 48,206–210.

[60] Tang, B; Yang, Z; Zhang, S. *J. Appl. Polym. Sci.*, 2012, 125, 1377–1381.

[61] Su, J-C; Liu, P-S. Energy Convers. *Manage.*, 2006, 47 (18-19), 3185–3191.

[62] Su, J-C; Liu, P-S. *Acta Polymerica Sinica*, 2007, 1 (2), 97-102.

[63] Meng, Q; Hu, J. *Sol. Energ. Mat. Sol. C.*, 2008, 92 (10), 1260–1268.

[64] Mondal, S; Hu, JL; Zhu, Y. *J. Membrane Sci.*, 2006, 280, 427-432.

[65] Xi, P; Xia, L; Fei, P; Zhang, D; Cheng, B. *Sol. Energ. Mat. Sol. C.*, 2012, 102, 36-43.

[66] Cao, Q; Liu, P. *Eur. Polym. J.*, 2006, 42 (11), 2931-2939.

[67] Chen, C; Wang, L; Huang, Y. *Polymer*, 2007, 48 (18), 5202–5207.

[68] Chen, C; Wang, L; Huang, Y. *AIChE J.*, 2009, 55, 820–827.

[69] Chen, C; Wang, L; Huang, Y. *Mater. Lett.*, 2009, 63 (5), 569-571.

[70] Chen, C; Wang, L; Huang, Y. *Appl. Energy*, 2011, 88 (9), 3133-3139.

[71] Seifpoor, M; Nouri, M; Mokhtari, J. *Fiber. Polym.*, 2011, 12 (6), 706-714.

[72] Nguyen, TTT; Park, JSJ. *Appl. Polym. Sci.*, 2011, 121, 3596–3603.

[73] Do, CV; Nguyen, TTT; Park, JS. *Sol. Energ. Mat. Sol. C.*, 2012, 104, 131-139.

[74] Chen, C; Zhao, Y; Liu, W. *Renew. Energ.*, 2013, 60, 222-225.

[75] Zhang, H; Yang, S; Liu, H; Fang, Y; Jiang, S; Guo, J; Gong, Y. *J. Appl. Polym. Sci.*, 2013,129, 1563–1568.

[76] Erkan, G. *RJTA*, 2004, 8 (2), 57-64.

[77] Vigo, TL; Frost, CM. United State Patent no. 4871615, Oct, 3, 1989.

[78] Shin, Y; Yoo, D-I; Son, K. *J. Appl. Polym. Sci.*, 2005, 96, 2005–2010.

[79] Ghosh, PS; Bhatkhande, P. *Int. J. Org. Chem.*, 2012, 2, 366-370.

[80] Karthikeyan, M; Ramachandran, T; Sundaram, OL. *S. J. Indus. Text.* 2014, 44 (1), 130-146.

[81] Hatfield, JC. United State Patent no 4,708,812, Nov. 24, 1987.

[82] Kumar, A; Kulkarni, PS; Samui, AB. *Cellulose*, 2014, 21, 685–696.

[83] Khudhair, AM; Farid, MM. *Energy Convers. Manage.*, 2004, 45, 263–275.

[84] Ip, KCW; Gates JR. Thermal storage for sustainable dwellings. In: Proceedings of International Conference, Sustainable Building. Maastricht, Netherlands, 2000.

[85] Li, H; Fang, G-Y. *Chem. Eng. Technol.*, 2010, 33, 1650–1654.

[86] Hu, S-G; Li, Q; Huang, S-L; Ding, Q-J. *Highway*, 2009, 07.

[87] Bentz, DP; Turpin, R. *Cement Concrete Comp.*, 2007, 29, 527–532.

[88] Raja, D; Balaji, HV; Karthikeyan M; Narmadha N; Visagavel K. *IJRET.*, 2014, 03 (11), 107-111.

[89] Madhumathi, AA; Sundarraja, BMC. *Int. J. Energ. Environ.*, 2012, 3 (5), 739-748.

[90] Sakulich, A; Bentz, DJ. *Mater. Civ. Eng.*, 2012, 24(8), 1034–1042.

[91] Sakulich, AR; Bentz, DP. *Construct. Build. Mater.*, 2012, 35, 483-490.

[92] Sarier, N; Onder, E. *Thermochim. Acta*, 2008, 475 (1-2), 15–21.

[93] Salyer, IO; Griffen, CW. United Stated Patent No. 4825939, Jan 01, 1989.

[94] Zhang, H; Guo, P; Fang, YP; Guo, J; Gong, YM. *Adv. Mater. Res.*, 2013, 821-822, 1164-1167.

INDEX

C

F

G

R